Reference and Practice Book

CORE-PLUS MATHEMATICS PROJECT

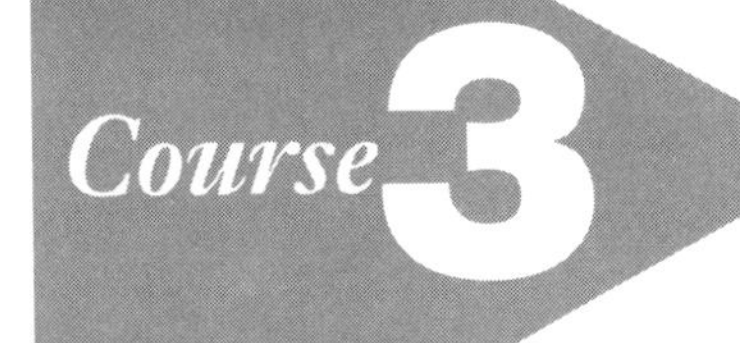

Contemporary Mathematics in Context

A Unified Approach

Arthur F. Coxford
James T. Fey
Christian R. Hirsch
Harold L. Schoen
Gail Burrill
Eric W. Hart
Ann E. Watkins
with the assistance of
Emma Ames, Robin Marcus,
Mary Jo Messenger, Jaruwan Sangtong,
Rebecca Walker, Edward Wall, and Marcia Weinhold

Chicago, Illinois

Project Directors
Arthur F. Coxford, University of Michigan
James T. Fey, University of Maryland
Christian R. Hirsch, Western Michigan University
Harold L. Schoen, University of Iowa

Senior Curriculum Developers
Gail Burrill, University of Wisconsin-Madison
Eric W. Hart, Western Michigan University
Brian A. Keller, Michigan State University
Ann E. Watkins, California State University, Northridge

Professional Development Coordinator
Beth Ritsema, Western Michigan University

Evaluation Coordinator
Steven W. Ziebarth, Western Michigan University

Project Collaborators
Emma Ames, Oakland Mills High School, Maryland
Robin Marcus, University of Maryland
Mary Jo Messenger, Howard County Public Schools, Maryland
Jaruwan Sangtong, University of Maryland
Rebecca Walker, Western Michigan University
Edward Wall, University of Michigan
Marcia Weinhold, Western Michigan University

Editorial and Production Assistants
James Laser, Western Michigan University
Kelly MacLean, Western Michigan University
Wendy Weaver, Western Michigan University

Everyday Learning Development Staff
Editorial: Anna Belluomini, Mary Cooney, Maureen Laude, Kathleen Ludwig, Ehrin Smith, Luke Zajac
Design/Production: Fran Brown, Elizabeth Gabbard, Jess Schaal, Silvana Valenzuela

Additional Credits
Design and Production: Lucy Lesiak Design

Photo Acknowledgments
Cover images: Images © 1997 Photodisc, Inc.
Cover Design: Oversat Paredes Design

This project was supported, in part,
by the
National Science Foundation
Opinions expressed are those of the authors
and not necessarily those of the Foundation

ISBN 1-57039-442-3

Everyday Learning Corporation
P.O. Box 81290
Chicago, IL 60681
www.everydaylearning.com

1 2 3 4 5 6 7 8 9 QW 07 06 05 04 03 02 01 0

Contents

Practicing for Standardized Tests 87

Solutions 129

Introduction

Course 2 of the *Contemporary Mathematics in Context (CMIC)* series introduced important ideas and problem solving techniques from algebra and functions, geometry, trigonometry, statistics, probability, and discrete mathematics. Many of those concepts and skills will reappear in Course 3 units. However, to make use of what you've learned, you may need periodic reminders of key ideas and practice with the skills that put those ideas to work. This *Reference and Practice (RAP)* book includes information and exercises that should be very helpful in reviewing and polishing the mathematics that you encountered in Course 2.

This book has three main sections: **Summary and Review of Course 2, Maintaining Concepts and Skills,** and **Practicing for Standardized Tests.**

The first section, Summary and Review of Course 2, contains summaries of key ideas developed in each of the four Course 2 mathematical strands:

- algebra and functions
- statistics and probability
- geometry and trigonometry
- discrete mathematics

The examples in this section illustrate application of the above strands to specific problems. Summaries of each topic are followed by a short review problem set to *Check Your Understanding*. It is a good idea to solve the *Check Your Understanding problems,* then check your solutions against the answer key at the back of the book.

The second section, Maintaining Concepts and Skills, contains twenty sets of review exercises from the various content strands mixed together as they might be in a cumulative examination or in real-life problem situation. These maintenance exercise sets draw primarily from material in Course 2 and should be used as periodic reviews to keep ideas from all strands fresh in your mind. Exercise Sets 1–10 review Course 2 and can be used at any time during the course. Since Exercise Sets 11–20 include some material from the first part of Course 3, they should be used any time during the second half of the course. Additional exercise sets for maintaining Course 3 concepts and skills are included in the *Teaching Resources* for Course 3.

The third section, Practicing for Standardized Tests, presents ten sets of questions that draw on all content strands. The questions are presented in the form of test items similar to how they often appear in standardized tests such as state assessment tests and college entrance examinations like the SAT and ACT. Use these test sets at any time during the school year to become familiar with the formats of standardized tests and to develop effective test-taking strategies for performing well on such tests.

Because you will probably use this book when studying outside of regular mathematics class sessions, answers to the problems are given at the end of the book. It is often possible to learn a lot by studying worked examples and working back from given answers to the required solution process. However, it's better to try to solve the problems on your own first, and then look at the answer key.

Summary and Review of Course 2

The four parts in this section of the Course 3 RAP book give brief summaries and illustrative examples for key topics in algebra and functions, statistics and probability, geometry and trigonometry, and discrete mathematics that you studied in Course 2. As you progress through *CMIC* Course 3, there may be activities or problems for which you need to use previously learned ideas that you don't completely remember. You can use this section to refresh your understanding of those ideas.

Within each topic summary you will find brief explanations of related concepts and methods, together with worked examples that are intended as a reference. These don't need to be studied from beginning to end. However, you should scan through this section so that you have an idea of what mathematics is reviewed and where various topics and subtopics are located. You can then refer to this section for specific information to help you when you need it.

1 Algebra and Functions

One of the main topics in high school mathematics is the study of relationships among quantitative variables. In Course 1, the quantitative relations you studied were primarily linear or exponential in form. In Course 2, the focus was primarily on power and quadratic relations.

1.1 Linear and Exponential Relations

Recall that *linear relations* are characterized by constant rates of change and graphs which are lines. A linear relation can be expressed by a linear equation: $y = a + bx$ where a is the y-intercept and b is the rate of change and slope of the graph. *Exponential relations* (growth or decay) are characterized by constant percent rates of change and nonlinear graphs which often rise or fall sharply. An exponential relation can be expressed by an exponential equation: $y = ab^x$ where a is the y-intercept and b is the growth (decay) factor.

EXAMPLE 1 ▶ A service station attendant earns a salary of \$280 for a 40-hour work week plus \$10.50 per hour overtime.

a. There is a linear relation between the earnings y and the number of hours of overtime worked per week x. The relation can be expressed by the linear equation: $y = 280 + 10.5x$, $x \geq 0$.

b. Linear relations may also be represented by *NOW-NEXT* equations: *NEXT* = *NOW* + b (start at *NOW* = a). The earnings-overtime relation is expressed by *NEXT* = *NOW* + 10.5 (start at 280).

c. The graph of the earnings-overtime relation is a line with a y-intercept of 280 and a slope of 10.5.

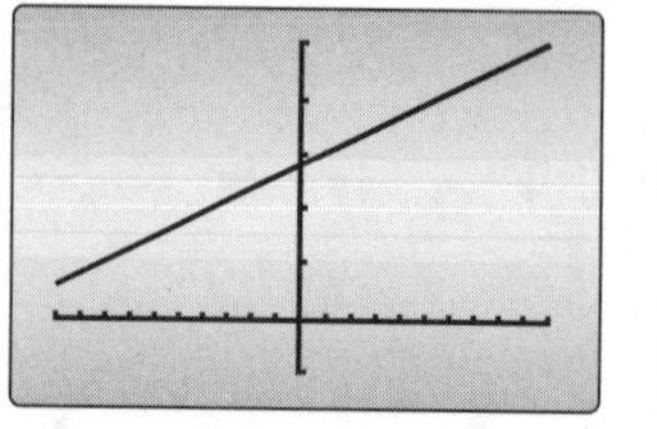

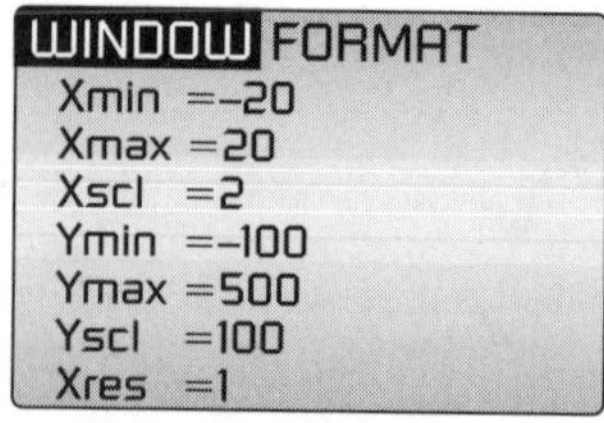

d. You can use a graph of the equation, a table of values, or symbolic reasoning to help you answer questions about the earnings-overtime relation. For example, if the attendant's earnings for a week were \$364, then the number of hours of overtime work can be found using any of these three methods as shown below. Zooming in on the graph and retracing will allow you to find a better approximation.

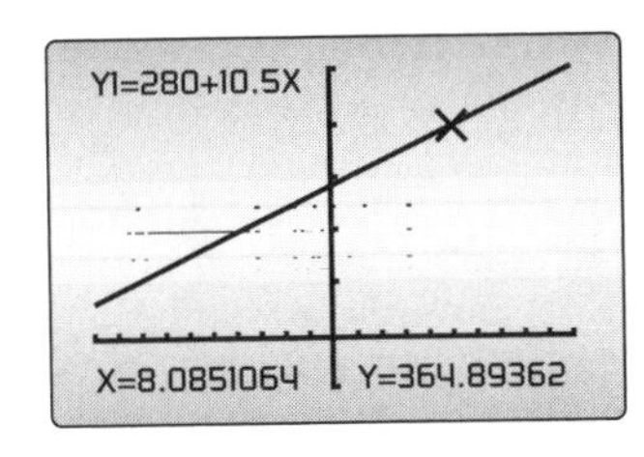

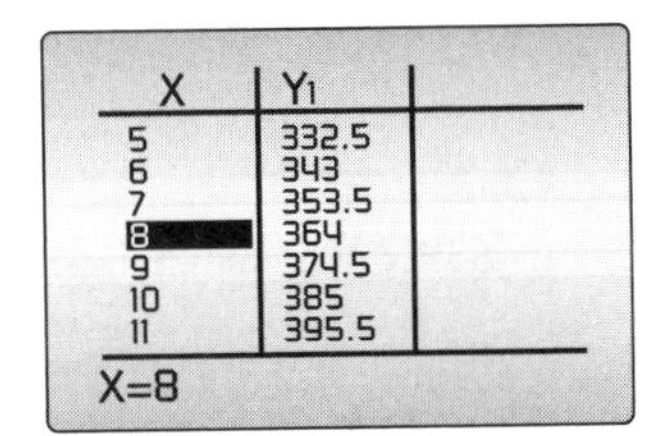

X	Y1
5	332.5
6	343
7	353.5
8	364
9	374.5
10	385
11	395.5

X=8

$$
\begin{aligned}
280 + 10.5x &= 364 \\
10.5x &= 84 \\
x &= 8
\end{aligned}
$$

EXAMPLE 2 ▶ Suppose a single bacterium lands in an open wound on your elbow and begins doubling every 30 seconds. How are the number of bacteria y and the number of 30-second time periods x related?

a. This is an exponential growth situation. The modeling equation is $y = 2^x$ for $x \geq 0$.

b. The same relation can be expressed in *NOW-NEXT* form: *NEXT* = 2 × *NOW* (start at 1).

c. The graph of $y = 2^x$ is an exponential curve as shown below. The y-intercept is 1.

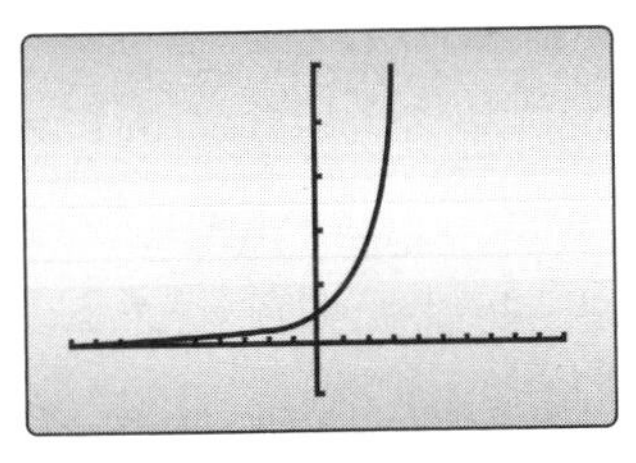

d. Tables can be especially helpful in reasoning about exponential growth and decay situations. For example, if there are 1,024 bacteria, how many hours have passed since the single bacterium landed on your elbow?

To begin, you need to find x so that $1{,}024 = 2^x$.

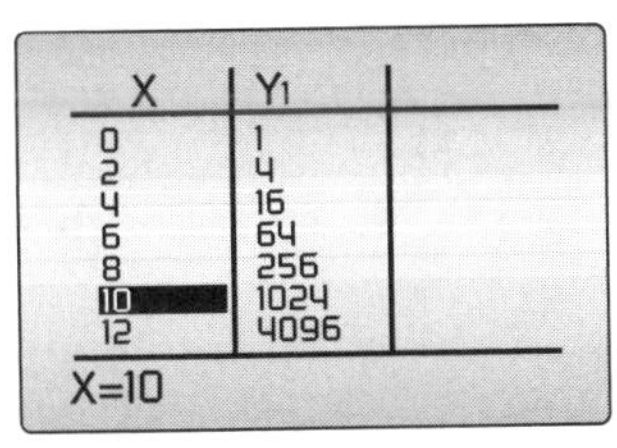

X	Y1
0	1
2	4
4	16
6	64
8	256
10	1024
12	4096

X=10

From the table it can be seen that $x = 10$. Since x represents 30-second intervals, 5 minutes will have passed.

e. If the infection started with not just one bacterium but with four bacteria, the modeling equation would be $y = 4(2^x)$.

Check Your Understanding 1.1

Solve the following problems to check your understanding of linear and exponential relations.

1. Find equations in $y = \ldots$ and *NOW-NEXT* forms that express the following problem conditions.

a. An airplane cruising at an altitude of 5,000 meters begins to descend at a rate of 250 meters per minute. Express the height of the descending airplane as a function of time.

b. The decision to cancel Saturday soccer games because of rain is made by the soccer club president and a three-member advisory group. Once the decision is made, the president and each member of the advisory group begin a calling tree by calling two other families apiece. Each family then calls two other families and so on. Express the number of calls as a function of the stage in the calling tree.

2. For each relation in Problem 1, sketch a graph.

3. Write and solve equations that match each of the following questions about the situations in Problem 1.

a. When does the plane in Part a reach an altitude of 1,000 meters?

b. At what stage of the calling tree in Part b will 512 families be notified of a cancellation?

4. The tables that follow show variables changing in predictable patterns.

I

x	0	1	2	3	4	5
y	2	6	18	54	162	486

II

x	0	1	2	3	4	5
y	2	6	10	14	18	22

a. Write *NOW-NEXT* equations that describe the pattern in each case.

b. Write $y = \ldots$ equations that describe the pattern in each table.

1.2 Linear Systems

Linear relations are often described with linear equations in the form $y = a + bx$. But in some linear situations where both variables are free to change subject to some constraints, the form $ax + by = c$ is commonly used. Each equation expresses a linear relation, the graphs are lines, the tables of corresponding values show the same pattern, but the symbolic forms are different.

Sometimes relations among variables in a situation can be expressed by a pair of linear equations rather than by a single such equation. A set of two or more linear equations is called a *linear system of equations*. The solution to such a system is the set of ordered pairs (x, y) that satisfy both equations in the system.

For example, the Grand Rapids White Caps are planning a promotional giveaway to motivate patrons to attend one of their baseball games. They plan to give an official little league baseball or bat to each of the first 2,000 children under 13 years of age who attend the designated game in July. They have $13,300 to spend. A ball costs $5, and a bat costs $8. How many balls and bats should the White Caps buy? You can answer this question by answering two related questions.

- What equations will summarize the two relationships described?
- Is there a solution that will satisfy both equations?

The following examples review the key concepts and skills required to answer these kinds of questions.

EXAMPLE 1 ▶ For the promotional giveaway by the Grand Rapids White Caps, the expression $x + y$ represents the total number of balls x and bats y bought for the promotion. Since they want to give away a total of 2,000 bats and balls, the equation $x + y = 2{,}000$ must be satisfied. The expression $5x + 8y$ represents the cost of x balls and y bats. If they use all the money, the equation $5x + 8y = 13{,}300$ must be satisfied.

EXAMPLE 2 ▶ Since $ax + by = c$ is equivalent to $y = -\frac{a}{b}x + \frac{c}{b}$, the graph of a linear relation $ax + by = c$ is always a straight line with y-intercept $\frac{c}{b}$ and slope $-\frac{a}{b}$. In Example 1, the equation $x + y = 2{,}000$ is equivalent to $y = -x + 2{,}000$, so the graph has slope -1 and y-intercept 2,000. The cost equation $5x + 8y = 13{,}300$ is equivalent to $y = -\frac{5}{8}x + \frac{13{,}300}{8}$, so its graph has slope $-\frac{5}{8}$ and y-intercept $\frac{13{,}300}{8}$. Since the lines have different slopes, they will intersect at a point whose coordinates satisfy both equations.

EXAMPLE 3 ▶ To determine the number of balls and bats to order that will use all of the budget, you need to solve the following system of linear equations:

$$\begin{aligned} x + y &= 2{,}000 \\ 5x + 8y &= 13{,}300. \end{aligned}$$

In Course 2, you examined five methods for solving linear systems. Methods 1, 2, and 5 often yield approximate answers. Methods 3 and 4 always produce exact answers.

Method 1: Graph the two equations and estimate the point of intersection of the resulting lines. This may be done by hand or using the trace capability of a graphing calculator. Graphs of the system

$$y = -x + 2{,}000 \text{ and } y = \frac{-5x + 13{,}300}{8}$$

are shown below.

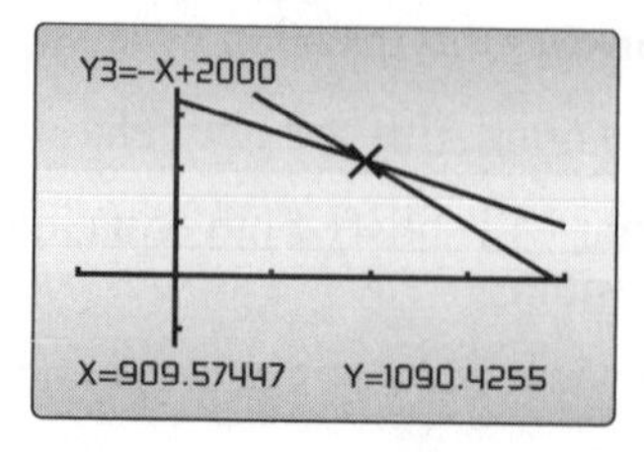

$-500 \leq x \leq 2{,}000$, scale: 500
$-1{,}000 \leq y \leq 2{,}000$, scale: 500

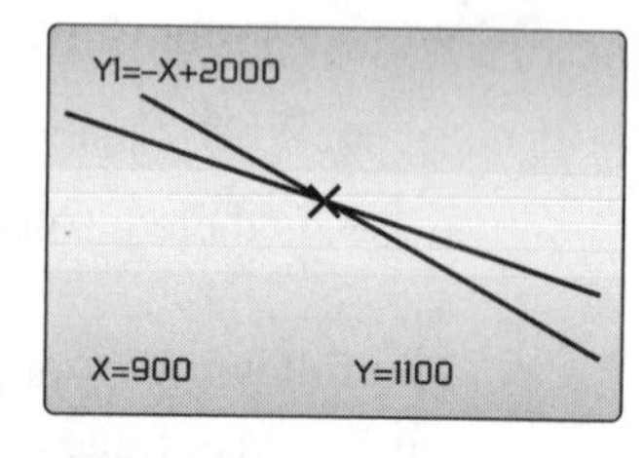

$899.9995 \leq x \leq 900.0005$
$1{,}099.9995 \leq y \leq 1{,}100.0005$

Zooming in and retracing allows you to improve estimates. The solution is $x = 900$ and $y = 1{,}100$. So, 900 balls and 1,100 bats should be ordered.

Method 2: Make a table of values for each of the equations and search for the x-value that gives the same y-value in both equations.

X	Y_1	Y_2
880	1120	1112.5
885	1115	1109.4
890	1110	1106.3
895	1105	1103.1
900	1100	1100
905	1095	1096.9
910	1090	1093.8

X=900

The table shows that when $x = 900$, the y-value of both equations is 1,100. The solution is (900, 1,100). A small change in either equation could make it very difficult to find the x-value that produces the same y-value for both equations.

Method 3: Solve each equation for y in terms of x and write an equation equating the expressions in x. Solve the resulting equation for x; then use that x-value to find the corresponding y-value.

Solving for y: $y = -x + 2{,}000$ and $y = -\frac{5}{8}x + \frac{13{,}300}{8}$

Equating expressions: $-x + 2{,}000 = -\frac{5}{8}x + \frac{13{,}300}{8}$

Solving for x: $\frac{3}{8}x = \frac{16{,}000}{8} - \frac{13{,}300}{8}$ or $3x = 2{,}700$

Thus $x = 900$. Substituting 900 in the first equation above gives $y = -900 + 2{,}000$ or $y = 1{,}100$.

Method 4: In this method, called the *linear-combination method*, you choose multipliers of the equations so that when the equations are added, the resulting sum equation has only one variable. For the system:

$$\begin{aligned} x + y &= 2{,}000 \\ 5x + 8y &= 13{,}300 \end{aligned}$$

multiplying the first equation by -5 gives the system:

$$\begin{aligned} -5x - 5y &= -10{,}000 \\ 5x + 8y &= 13{,}300. \end{aligned}$$

Add the two equations to get $3y = 3{,}300$, from which it follows that $y = 1{,}100$ and $x = 900$.

Method 5: In this matrix method, you write the system as a matrix equation and then multiply both sides by the inverse of the coefficient matrix. The matrix representation of the system:

$$\begin{aligned} x + y &= 2{,}000 \\ 5x + 8y &= 13{,}300 \end{aligned}$$

is $$\begin{bmatrix} 1 & 1 \\ 5 & 8 \end{bmatrix} \begin{bmatrix} x \\ y \end{bmatrix} = \begin{bmatrix} 2{,}000 \\ 13{,}300 \end{bmatrix}$$

Since the inverse of the coefficient matrix is $\frac{1}{3}\begin{bmatrix} 8 & -1 \\ -5 & 1 \end{bmatrix}$,

the solution is

$$\frac{1}{3}\begin{bmatrix} 8 & -1 \\ -5 & 1 \end{bmatrix}\begin{bmatrix} 1 & 1 \\ 5 & 8 \end{bmatrix}\begin{bmatrix} x \\ y \end{bmatrix} = \frac{1}{3}\begin{bmatrix} 8 & -1 \\ -5 & 1 \end{bmatrix}\begin{bmatrix} 2{,}000 \\ 13{,}300 \end{bmatrix}$$

or

$$\begin{bmatrix} x \\ y \end{bmatrix} = \begin{bmatrix} 900 \\ 1{,}100 \end{bmatrix}$$

Check Your Understanding 1.2

Solve the following problems to check your understanding of methods of solving linear systems.

1. Kelly burns 7.3 calories per minute when using a rowing machine and 5.7 calories per minute when using a stair machine. She plans to exercise for 45 minutes and wants to burn 300 calories. How many minutes should she spend on each machine?

a. What system of equations can be used to model this situation?

b. Solve the linear system in Part a in three different ways.

2. The Beaker Street Flea Market charges residents of Beaker Street a base fee of \$5 plus \$4 per hour to rent a display area. For non-residents, the market charges \$8 per hour rent but no base fee for the privilege of displaying wares. The market opens at 8:00 A.M. Find the rental time for which the display costs are the same for the two groups. What is that cost?

3. Find exact solutions to each system of equations. Use at least two different methods.

a. $x + y = 6$
$x - y = 4$

b. $4y = x + 2$
$x + y = 7$

c. $x + y = 4$
$y - x = 2$

d. $3x + 2y = 4$
$2x - 3y = 7$

e. $3x + y = 3$
$x - 4y = 1$

f. $x = -5 - 4y$
$3x + y = -4$

1.3 Power Relations

In addition to the linear and exponential families of functions, another important family of functions is *power functions*. The standard form of symbolic rules for all power models is $y = ax^p$. When $p > 0$, as in $y = -5x^2$ or $y = 4x^3$, the power model represents *direct variation*. When $p < 0$, as in $y = 2x^{-1} = \frac{2}{x}$ or $y = 3x^{-2} = \frac{3}{x^2}$, the power model represents *inverse variation*. In the case of inverse variation, the two quantities have a constant nonzero product. For example, if $y = \frac{2}{x}$, then $xy = 2$. Similarly, if $y = \frac{3}{x^2}$, $x^2y = 3$.

More generally, if $y = ax^p$, *y varies directly as the pth power of x.* If $x^p y = a$, $x \neq 0$, or equivalently $y = \frac{a}{x^p}$, *y varies inversely as the pth power of x.* The constant a is called the *constant of variation.*

EXAMPLE 1 ▶ The power generated by a windmill is directly proportional to the cube of the wind speed. If a 10 mph wind generates 150 watts of power, how many watts will a 20 mph wind generate? If W represents the wind speed and P represents the power generated, then $P = aW^3$. So $150 = a(10^3)$ and $a = \frac{150}{1,000} = 0.15$. In the case of a 20 mph wind, $P = 0.15(20^3) = 1,200$ watts.

EXAMPLE 2 ▶ The number of tomato plants in a row in Jim's garden varies inversely as the space between the plants. If the plants are spaced 15 inches apart, 60 plants fit in a row. How many plants can be planted in one row if the space between plants is 25 inches? If x represents the space between plants and y represents the number of plants in a row, then $y = \frac{a}{x}$ or $xy = a$. Hence, $15 \times 60 = a$ or $a = 900$. So, if the space between plants is 25 inches, $y = \frac{900}{25} = 36$. So 36 plants can be planted in one row.

EXAMPLE 3 ▶ The graph of $y = \frac{a}{x}$ for $a > 0$ is shown at the right. Notice that the graph approaches both axes, but it reaches neither. This graph is called a *hyperbola.*

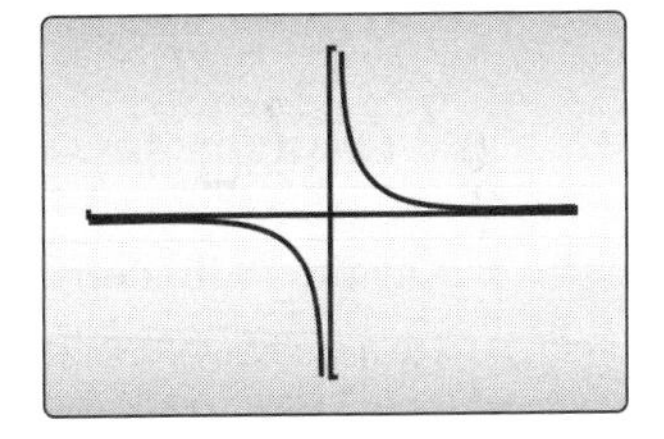

If $a < 0$, the graph of the function $y = \frac{a}{x}$ shown at the right is also a hyperbola. However it is the reflection across the y-axis of the first graph.

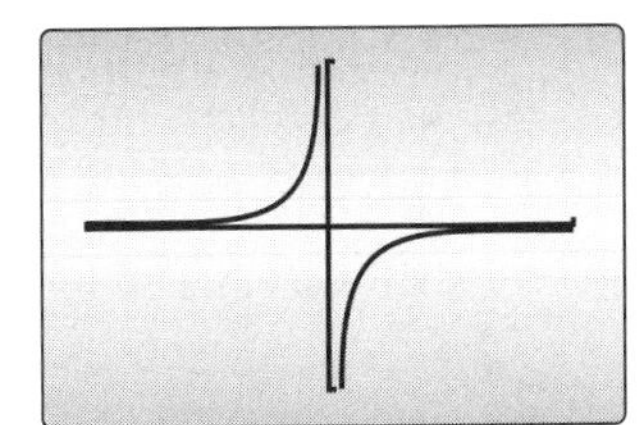

Check Your Understanding 1.3

Check your understanding of power functions and the related ideas of direct and inverse variation by solving the following problems.

1. Model each of the following situations with an equation.

a. The strength of an animal's bones varies directly with the square of its height.

b. The weight of an animal varies directly with the cube of its height.

c. The loudness of a sound from a stereo speaker varies inversely as the square of the distance from the speaker.

2. When a wire, such as a guitar string, is plucked, it vibrates at a frequency that depends on the tension and length of the wire. For any fixed tension, the frequency varies inversely as the length of the wire. If a wire that is 400 centimeters long vibrates 160 times per second, how long should a wire be to vibrate 240 times per second (at the same tension)?

3. The frequency f of vibration of a wire (of uniform length and tension) varies inversely as the square root of its weight w. Express this relation by an equation, using k to represent the constant of variation. Given that k is 508, what is f when w is 4?

1.4 Quadratic Relations

Quadratic functions with rules of the form $y = ax^2 + bx + c$ can be thought of as a sum of a power function and a linear function. Quadratic functions are useful in modeling business profits, stopping distances, and paths of balls thrown or kicked in the air. For example, the punter on a local high school football team typically kicks the ball at a point 2 feet off the ground, with an initial upward velocity of 60 feet per second. The height h of the ball above the ground in feet is a function of time t since the ball was kicked in seconds and can be modeled by the quadratic function:

$$h = -16t^2 + 60t + 2.$$

You have continued to develop skill in the use of algebraic ideas to answer questions like these:

- About how much time does the kick receiver have to wait to receive the ball?
- What is the maximum height of the kicked ball?
- How would the flight of the kicked ball change if the initial upward velocity were increased or decreased?

The answers to questions such as these may be found using graphical, tabular, or symbolic reasoning strategies.

EXAMPLE 1 ▶ Estimate the elapsed time t before the kicked ball hits the ground. You can use a graph to estimate this time. Since the ball hits the ground when the height is 0, trace the graph of $y = -16t^2 + 60t + 2$ in search of a y-value that is approximately 0. The time is about 3.8 seconds.

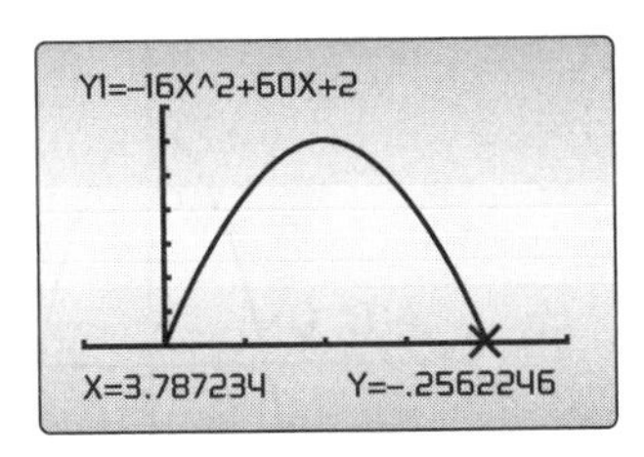

A tabular approach might begin with the table on the left below.

X	Y1
3.6	10.64
3.65	7.84
3.7	4.96
3.75	2
3.8	-1.04
3.85	-4.16
3.9	-7.36

X=3.8

X	Y1
3.78	.1856
3.781	.12462
3.782	.06362
3.783	.00258
3.784	-.0585
3.785	-.1196
3.786	-.1807

X=3.783

Y_1 is 0 for a value of x between 3.75 and 3.80. Changing the step for x to 0.001 suggests that a better estimate is 3.783.

EXAMPLE 2 ▶ Symbolic reasoning can be used to solve quadratic equations like $-16x^2 + 2 = 0$ or $16x^2 - 2 = 47$. For example,

$-16x^2 + 2 = 0$ is equivalent to $2 = 16x^2$. (Add $16x^2$ to both sides.)

$\frac{2}{16} = x^2$ (Divide both sides by 16.)

$\pm\sqrt{\frac{1}{8}} = x$ (Definition of x^2)

$16x^2 - 2 = 47$ is equivalent to $16x^2 = 49$. (Add 2 to both sides.)

$x^2 = \frac{49}{16}$ (Divide both sides by 16.)

$x = \pm\sqrt{\frac{49}{16}}$ (Definition of x^2)

$x = \pm\frac{7}{4}$ (Definition of a square root)

Note: In Course 3, you will develop a formal method for solving the general quadratic equation: $ax^2 + bx + c = 0$.

EXAMPLE 3 ▶ The coefficients of a quadratic function rule determine whether its graph opens up or down, its axis of symmetry, and its x- and y-intercepts.

$y = -2x^2 + 18$

y
12
(−3, 0)
(3, 0)
−6
6 x
−12

axis of symmetry: $x = 0$

$y = x^2 - 2x - 8$

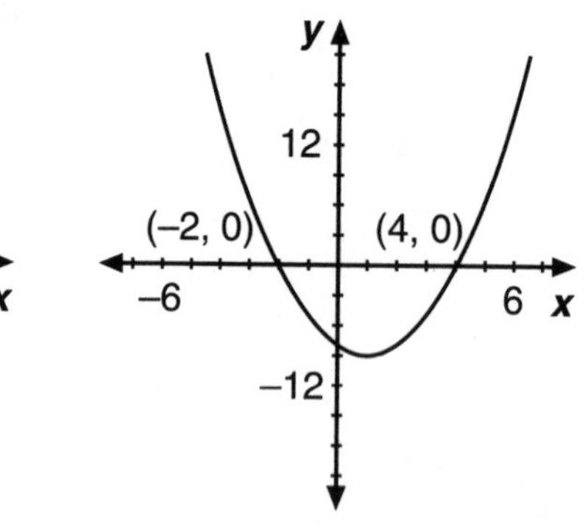

axis of symmetry: $x = 1$

In general, for the quadratic function $y = ax^2 + bx + c$:

a. The graph opens upward when $a > 0$ and downward when $a < 0$;

b. The y-intercept is $(0, c)$;

c. The axis of symmetry is the line $x = \frac{-b}{2a}$ or $-\frac{b}{2a}$;

d. The x-intercepts are equidistant from the axis of symmetry and can be approximated using a table or graph.

Check Your Understanding 1.4

Check your understanding of quadratic functions by completing the following tasks.

1. Estimate the solution of each quadratic equation using calculator-produced tables or graphs.

a. $x^2 + 4x - 12 = 0$ **b.** $4x^2 - x - 5 = 0$

c. $3x^2 - 12x + 12 = 0$ **d.** $3x^2 - 2x + 8 = 0$

2. A rectangular strip of asphalt paving is 5 meters longer than it is wide. Its area is 300 square meters. What are the length and the width of the strip?

3. Use symbolic reasoning to solve each quadratic equation. Leave solutions in radical form.

a. $x^2 - 16 = 0$ **b.** $x^2 - 27 = 0$ **c.** $x^2 - 17 = 19$

d. $-x^2 - 18 = 30$ **e.** $3x^2 - 8 = 16$ **f.** $2x^2 - 12 = -3$

4. For each quadratic function, describe the nature of its graph, including the direction it opens, its axis of symmetry, and its x- and y-intercepts.

a. $y = 2x^2 - 5x - 12$ **b.** $y = -2x^2 + 12$

c. $y = 3x^2 - x - 2$ **d.** $y = -x^2 + 2x + 1$

1.5 Powers, Roots, and Radicals

Exponents can sometimes provide an economical way to rewrite products involving the same factor. The important idea to keep in mind is the meaning of an exponent.

If x is a real number and m is an integer greater than 1, then
$x^m = \underbrace{x \cdot x \cdot x \cdot \ldots \cdot x}_{m \text{ factors}}$. Note, $x^1 = x$.

In many situations it is informative to write a given exponential expression in an equivalent form.

EXAMPLE 1 ▶ For the *product of two powers* with the same base: $x^m \cdot x^n = x^{m+n}$. For example, $7^3 \cdot 7^5 = 7^{3+5} = 7^8$. When there are more than two factors, the rule extends in the obvious way: $3^2 \cdot 3 \cdot 3^4 \cdot 3^3 = 3^{2+1+4+3} = 3^{10}$.

EXAMPLE 2 ▶ For the *power of a power:* $(x^m)^n = x^{m \cdot n}$.

For example, $(4^3)^5 = (4)^{3 \cdot 5} = 4^{15}$.

EXAMPLE 3 ▶ For the *power of a product:* $(x \cdot y)^m = x^m \cdot y^m$.

For example, $(3 \cdot 6)^5 = 3^5 \cdot 6^5$.

EXAMPLE 4 ▶ For a power with a *negative integer exponent:* $x^{-m} = \frac{1}{x^m}$, $x \neq 0$, $m > 0$.

In particular, $x^{-1} = \frac{1}{x}$.

For example, $10^{-1} = \frac{1}{10}$ and $2^{-5} = \frac{1}{2^5} = \frac{1}{32}$.

EXAMPLE 5 ▶ For the *quotient of powers* with the same base: $\frac{x^m}{x^n} = x^{m-n}$, $x \neq 0$.

For example, $\frac{10^4}{10^6} = 10^{4-6} = 10^{-2}$ or $\frac{1}{10^2}$.

Note that $x^0 = 1$ since $1 = \frac{x^1}{x^1} = x^{1-1} = x^0$.

Exponentiation, raising a number to a power n, requires you to find the product using the number as a factor n times. If you know that a number is the product of n equal factors, then finding the value of that factor is *finding the nth root* of the number. For example, the third (or cube) root of 64 is 4 because $4^3 = 64$. Thus, the solution to the equation $x^3 = 64$ is $x = 4$.

EXAMPLE 6 ▶ Finding the square root of 48 is equivalent to solving the equation $x^2 = 48$. Symbolically, $x = \pm\sqrt{48}$. The symbol, $\sqrt{\ }$, is called the *radical symbol* and indicates that you should find the square or second root of the number. For example, $\sqrt{48}$ is about 6.928 since $6.928^2 \approx 48$. In general, the symbol $\sqrt[n]{\ }$ indicates the nth root rather than the second root. For example, $\sqrt[3]{64} = 4$ since $4^3 = 64$.

EXAMPLE 7 ▶ The *root of a product* is the product of the roots of the factors: $\sqrt{x \cdot y \cdot z} = \sqrt{x} \cdot \sqrt{y} \cdot \sqrt{z}$, $x, y, z \geq 0$.

For example,
$$\begin{aligned} \sqrt{48} &= \sqrt{16 \cdot 3} \\ &= \sqrt{16} \cdot \sqrt{3} \\ &= 4 \cdot \sqrt{3} \text{ or } 4\sqrt{3} \end{aligned}$$

The last expression is sometimes referred to as the *simplified form* of the radical expression $\sqrt{48}$. Note that $\sqrt[3]{48}$ can be rewritten as $2\sqrt[3]{6}$ since $\sqrt[3]{48} = \sqrt[3]{8} \cdot \sqrt[3]{6}$ and $\sqrt[3]{8} = 2$.

EXAMPLE 8 ▶ The exponential form of a radical uses *fractional exponents.* $\sqrt{48}$ can be written equivalently as $48^{\frac{1}{2}}$ or $48^{0.5}$. The cube root of 48 is written $\sqrt[3]{48}$ or $48^{\frac{1}{3}}$.

Check Your Understanding 1.5

Check your understanding of exponents and radicals by completing the following review exercises.

1. Write each exponential expression in an equivalent and simpler form. Variables represent nonzero numbers.

a. $(x^2)^3$ **b.** $(a^3)^4$ **c.** $(x^2)^{-3}$ **d.** $\frac{c^{12}}{c^6}$

e. $\frac{3^5}{3^2}$ **f.** $\frac{3^2}{3^5}$ **g.** x^4x^6 **h.** $\frac{a^3b^2}{ab^5}$

2. Rewrite each radical expression in an equivalent, simpler form. Variables under a radical represent positive numbers.

a. $\sqrt{24}$ **b.** $\sqrt{243}$ **c.** $\sqrt[3]{243}$ **d.** $\sqrt[3]{32}$

e. $\sqrt{8a^3b^5}$ **f.** $\sqrt[3]{64x^5y^6}$ **g.** $72^{\frac{1}{2}}$ **h.** $72^{\frac{1}{2}}$

3. Write an equivalent expression for each given expression. No variable is equal to zero.

a. t^{-8} **b.** $\left(\frac{1}{2}\right)^{-4}$ **c.** $(-2)^{-3}$ **d.** $(-2)^{-4}$

e. $\frac{a^7 \cdot a^9}{a^6}$ **f.** $\frac{(c^2 \cdot c^3)^4}{c^7}$ **g.** $\frac{x^{-10} \cdot x^8}{x^{-2}}$ **h.** $\frac{b^3 \cdot b^{-9}}{b^{-3}}$

2 Statistics and Probability

2.1 Correlation and Regression

In your study of the first two *CMIC* courses, you saw that a scatterplot is an effective way to display the relation between two quantitative variables. Two variables often are *associated* linearly, meaning the points cluster, tightly or loosely, about a line. Linear association may be determined by visual inspection of a scatterplot. The strength of the linear association can be measured by Pearson's *correlation* coefficient. *Least squares regression* provides a method for finding the line that best fits the linear data pattern.

EXAMPLE 1 ▶ Scatterplot I on the left below displays data that exhibit a *strong positive* linear association. Scatterplot II on the right below shows data that have a *strong negative* linear association.

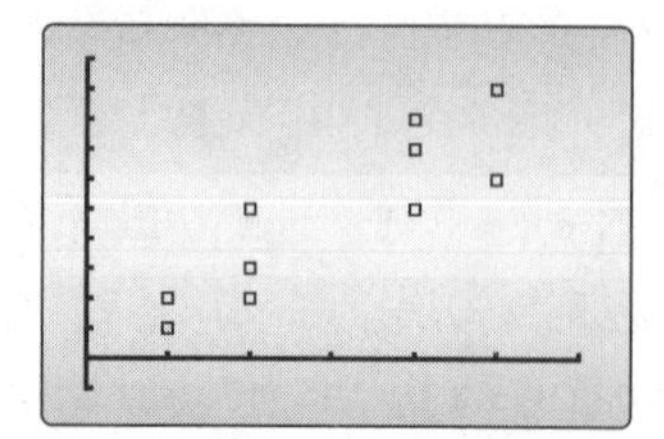

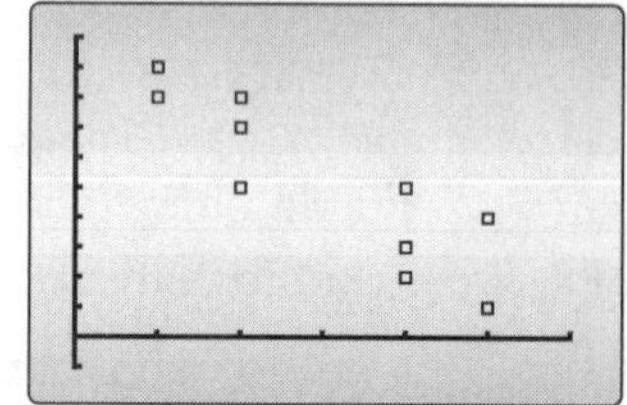

In general, if data cluster about a line with positive slope, the association is positive; if they cluster about a line with negative slope, the association is negative. The strength of the association depends on how tightly the data cluster about a line. The correlation coefficient will always be between −1 and 1, inclusive. The closer the absolute value of the correlation coefficient is to 1, the stronger the association.

EXAMPLE 2 ▶ The strength of the linear association of two variables is measured by Pearson's *correlation coefficient, r*. The equation used to compute r is

$$r = \frac{\Sigma(x - \bar{x})(y - \bar{y})}{\sqrt{\Sigma (x - \bar{x})^2}\sqrt{\Sigma (y - \bar{y})^2}}$$

Technology is usually used to do the actual computation of r. Consider the data corresponding to the scatterplots in Example 1 and given in the tables at the top of page 21.

Plot I

x	1	1	2	2	2	4	4	4	5	5
y	1	2	2	3	5	5	7	8	6	9

For the (x, y) data, r is 0.88.

Plot II

x	1	1	2	2	2	4	4	4	5	5
z	9	8	8	7	5	5	3	2	4	1

For the (x, z) data, the correlation coefficient is -0.88.

EXAMPLE 3 ▶ *Influential points* can change the value of r dramatically. If the point (6, 1) is included in the set of (x, y) data, the correlation coefficient changes significantly. The new correlation coefficient is about 0.48 or about half of the previous value. Because of this, you should always examine a scatterplot to determine if one or two *outlying* points are influencing the correlation coefficient.

EXAMPLE 4 ▶ The *least squares regression line* is the line of best fit for a set of data that follows a linear pattern. It is considered the best fit because the sum of the squares of the *residuals* for the least squares regression line is smaller than for any other line. A *residual* is the difference between the y-value observed in the data and the y-value predicted by the regression equation.

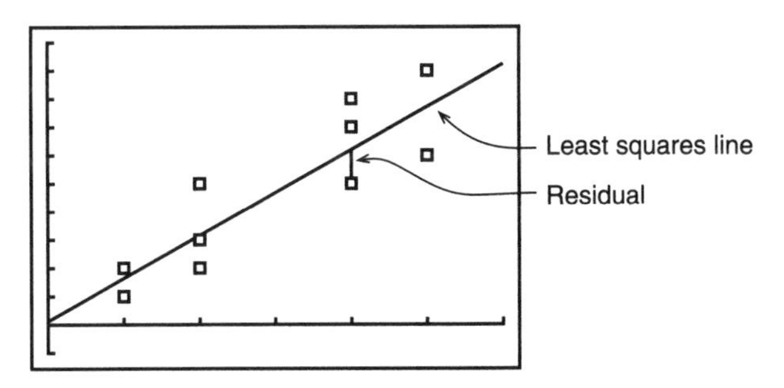

For example, the least squares regression equation for the (x, y) data in Example 2 is $y = 1.545x + 0.164$. For the point (5, 6), the residual is *observed – predicted* which is $6 - (1.545 \cdot 5 + 0.164)$ or -1.889.

EXAMPLE 5 ▶ The mean of the x-values of the (x, y) data in Example 2 is 3; the mean of the y-values is 4.8. The point with these means as coordinates is the *centroid* of the scatterplot. The regression line always contains the centroid as the following computation illustrates: $1.545 \cdot 3 + 0.164 = 4.799$. The slight difference here is due to rounding of the x-coefficient and the constant term in the regression equation.

Check Your Understanding 2.1

Solve the following problems to check your understanding of linear association and least squares regression lines.

1. The following table provides the number of calories and sodium content (in mg) for selected fast foods.

Calories	290	420	450	550	300	200	250	325
Sodium (mg)	860	1,190	740	480	530	360	575	640

a. Make a scatterplot of the data. Describe the strength of its linear association.

b. Calculate the correlation coefficient.

c. Add the point (710, 1400) to the scatterplot. Predict the effect of this point on the correlation coefficient. Is your prediction correct?

d. Show that the regression line goes through the centroid of the data. (Include the point (710, 1400) in your data set.)

2. Use the data from Problem 1, Part c, but delete the point (550, 480).

a. How does the deletion of this point affect the linear association?

b. How is this reflected in the change in the correlation coefficient?

c. Both (550, 480) and (710, 1400) are influential points in the data. How could you determine this by examining the scatterplot? How does each point influence the correlation coefficient?

3. Using the original data from Problem 1:

a. Calculate and graph the least squares regression equation. Interpret the slope of the line in the context of the data.

b. Calculate the residual for the point (550, 480).

2.2 Waiting-Time Distributions and Expected Value

In Course 1, you learned how to estimate probabilities using simulation. For some probabilistic situations, such as rolling a die or flipping coins, it is possible to calculate exact probabilities. In Course 2, you constructed *probability distributions* which show you all possible outcomes of a probabilistic situation along with their respective probabilities. For example, if you roll a die twice, the probability distribution of the sum of the two numbers on the two rolls is given in the following table.

Sum: x	2	3	4	5	6	7	8	9	10	11	12
Probability: $P(x)$	$\frac{1}{36}$	$\frac{2}{36}$	$\frac{3}{36}$	$\frac{4}{36}$	$\frac{5}{36}$	$\frac{6}{36}$	$\frac{5}{36}$	$\frac{4}{36}$	$\frac{3}{36}$	$\frac{2}{36}$	$\frac{1}{36}$

EXAMPLE 1 ▶ In constructing a probability distribution, sometimes you need to know the probability that two events both happen. For example, if you roll a die twice, what is the probability that the sum is 12? The only way to get this sum is to roll a 6 both times. Since the probability of a 6 on the first roll is $\frac{1}{6}$ and the probability of a 6 on the second roll is $\frac{1}{6}$, the probability of two successive 6s is $\left(\frac{1}{6}\right)\left(\frac{1}{6}\right) = \frac{1}{36}$. This multiplication rule works only because the rolls are *independent*; that is, the probabilities for the second roll are not affected by the result of the first roll. If two events, A and B, are independent, you find the probability that they both happen by using the *Multiplication Rule for Independent Events*: $P(A \text{ and } B) = P(A) \cdot P(B)$.

EXAMPLE 2 ▶ Suppose you have a bag with 5 red balls and 3 green balls. If you draw 2 balls, what is the probability they are both red? The answer depends on whether you replace the first ball before drawing the second. If you do, the probability of getting a red ball on the first draw and a red ball on the second draw is $\frac{5}{8} \times \frac{5}{8} = \frac{25}{64}$. The two events are independent since the probability of getting a red ball on the second draw remains $\frac{5}{8}$. However, if you don't replace the first ball before drawing the second, the probability is $\frac{5}{8} \times \frac{4}{7} = \frac{20}{56}$. With one red ball gone, the probability the second ball will be red drops to 4 out of 7. The two trials are *dependent* since the result of the first draw

changes the probabilities for the second draw. If two events, A and B, are dependent, you find the probability that they both happen by using the *General Multiplication Rule:*
$P(A \text{ and } B) = P(A \text{ happens}) \cdot P(B \text{ happens given that } A \text{ happens})$.

EXAMPLE 3 ▶ Suppose that Big Burger randomly selects one-fourth of the purchasers of their double-patty Meat Extravaganza to receive a free drink. Buying burgers until you get a free drink is an example of a *waiting-time* situation. The probability distribution for this waiting-time situation is given by the table below.

Number of Purchases to Get First Free Drink: x	1	2	3	4	5	...
Probability: $P(x)$	$\frac{1}{4}$	$\frac{3}{4} \times \frac{1}{4}$	$\left(\frac{3}{4}\right)^2 \times \frac{1}{4}$	$\left(\frac{3}{4}\right)^3 \times \frac{1}{4}$	$\left(\frac{3}{4}\right)^4 \times \frac{1}{4}$	...

For example, the probability of getting a free drink for the first time on the third purchase is $\frac{3}{4} \times \frac{3}{4} \times \frac{1}{4}$ because you must "fail," "fail," and then "succeed." In a waiting-time distribution, the *average (or expected) waiting time* is $\frac{1}{p}$ where p is the probability of the event occurring. In this situation, the average waiting time to get a free drink is $\frac{1}{p} = \frac{1}{\frac{1}{4}} = 4$ purchases.

EXAMPLE 4 ▶ The *expected value* (*EV*) of a probability distribution is the mean of the distribution. EV is computed using the formula: $\text{EV} = \Sigma\, x \cdot P(x)$ for all pairs $(x, P(x))$. For example, suppose that you throw a die and get \$2 if you throw a 3 or 6; get \$0.50 for a 1; and *pay* \$1 if you throw a 2, 4, or 5. The expected value, the average amount of money you get per throw in the long run, is $2\left(\frac{2}{6}\right) + 0.5\left(\frac{1}{6}\right) - 1\left(\frac{3}{6}\right) = 0.25$. To make this a *fair game*, you should pay \$0.25 to play!

EXAMPLE 5 ▶ In a waiting-time distribution, a *rare event* is an event that falls in the upper 5% of the distribution. For example, in the waiting-time distribution for getting a free drink, having to buy 11 or more burgers before getting a free drink has a probability of $1 - [P(1) + P(2) + \ldots + P(10)]$, or about 0.056, while having to buy 12 or more burgers has a probability of about 0.042. So, having to buy 12 burgers (or more) before getting a free drink is a rare event.

Check Your Understanding 2.2

Check your understanding of waiting-time distributions, probabilities of independent events, and expected value by solving the following problems.

1. Find each of the following probabilities.

a. Drawing two balls without replacement from a bag of 15 black balls and 20 red balls and getting two black balls

b. Drawing two cards without replacement from a deck of 52 playing cards and getting two spades

c. Drawing two cards with replacement from a deck of 52 playing cards and getting two spades

d. Drawing two cards without replacement from a deck of 52 playing cards and getting the ace of hearts on the first draw and the 2 of diamonds on the second draw

e. Flipping a coin five times and getting three heads followed by two tails

2. A spinner spins on a circular region that is divided into four colors: $\frac{1}{3}$ yellow, $\frac{1}{6}$ red, $\frac{3}{16}$ blue, and $\frac{5}{16}$ green. Find the expected value of a spin if you get $1 for yellow and $10 for red, but *pay* $3 for blue and $4 for green.

3. A cereal manufacturer places a prize in 20% of the boxes of Breakfast Munchies.

a. Construct the waiting-time distribution for the number of boxes purchased to get the prize.

b. Find the expected value of the distribution.

c. Is it a rare event to have to wait until the ninth purchase of a box of Munchies to get a prize?

3 Geometry and Trigonometry

Geometry provides you with a language for the description and analysis of shapes and their properties. You have seen how the characteristics of a shape, such as its rigidity or flexibility, play a role in how objects of that shape function in the real world. The introduction of coordinates provides another way of representing and reasoning about geometric figures and transformations of figures in a plane. Coordinates enable you to think about geometric ideas such as parallelism, algebraically in terms of slopes of lines; they also allow you to think of function graphs geometrically in terms of their position in the plane.

The introduction of the trigonometric ratios—sine, cosine, and tangent—for right triangles provides a connection between angle measure and linear measure and methods for calculating distances that could not otherwise be measured. Trigonometric ratios also provide the foundation for describing circular motion and for modeling periodic phenomena.

The following sections review the key concepts and skills related to coordinate geometry, transformations, and trigonometry.

3.1 Coordinate Geometry

A point in a plane can be located with respect to two perpendicular lines by using a pair of real numbers. The lines are called the *axes*, the point of intersection of the axes is called the *origin*, and the pair of real numbers locating the point is called the *coordinates*, (x, y), of the point.

EXAMPLE 1 ▶ Two points uniquely determine a line. Thus, there is a unique line containing $A(2, -5)$ and $B(-3, 6)$. Recall that the *slope, m,* of that line is found by calculating the ratio of the difference in the y-coordinates to the difference in the x-coordinates, $m = \frac{y_1 - y_2}{x_1 - x_2}$. In particular, the slope of the line containing points A and B is $\frac{-5-6}{2-(-3)}$ or $\frac{-11}{5}$. Verify that the slope of the line containing points $C(2, -4)$ and $D(3, 2)$ is 6.

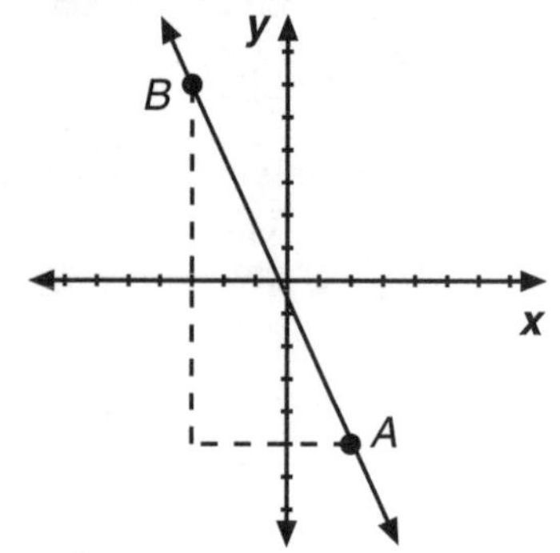

EXAMPLE 2 ▶ The length of the segment joining points $A(x_1, y_1)$ and $B(x_2, y_2)$ can be found by using the *distance formula:*

$d = \sqrt{(x_1 - x_2)^2 + (y_1 - y_2)^2}$. For example, if the endpoints of $\overline{AB}$ are $A(2, -5)$ and $B(-3, 6)$, then

$AB = \sqrt{(2-(-3))^2 + (-5-6)^2} = \sqrt{(5)^2 + (-11)^2} = \sqrt{146} \approx 12.1.$

EXAMPLE 3 ▶ The coordinates of the *midpoint* of $\overline{AB}$ with $A(2, -5)$ and $B(-3, 6)$ are $\left(\frac{2+(-3)}{2}, \frac{-5+6}{2}\right)$ or $\left(\frac{-1}{2}, \frac{1}{2}\right)$. The general formula for the coordinates of the midpoint of $\overline{AB}$ with $A(x_1, y_1)$ and $B(x_2, y_2)$ is $\left(\frac{x_1+x_2}{2}, \frac{y_1+y_2}{2}\right)$.

EXAMPLE 4 ▶ The slopes of *parallel lines* are identical. The slopes of *perpendicular lines* are negative reciprocals of each other. For example the slope of $\overleftrightarrow{AB}$ in Example 1 is $-\frac{11}{5}$. Thus, the slope of any line parallel to $\overleftrightarrow{AB}$ is $-\frac{11}{5}$ and the slope of any line perpendicular to $\overleftrightarrow{AB}$ is $\frac{5}{11}$. Note that $-\frac{11}{5} \cdot \frac{5}{11} = -1$.

EXAMPLE 5 ▶ The coordinate representation of points also permits the use of matrices to represent polygons. For example, $\triangle ABC$ determined by $A(1, -2)$, $B(3, 1)$, and $C(2, 2)$ can be represented by the matrix

$$\triangle ABC = \begin{bmatrix} 1 & 3 & 2 \\ -2 & 1 & 2 \end{bmatrix}$$

in which columns of the matrix are the coordinates of the vertices.

Check Your Understanding 3.1

Check your understanding of the basic ideas of coordinate geometry by completing the following problems.

1. Find the slope and the equation of the line containing each pair of points.

a. $A(2, 3)$ and $B(-4, 5)$ **b.** $R(-4, -3)$ and $S(5, 1)$

2. Consider $\triangle PQR = \begin{bmatrix} -2 & -5 & -3 \\ 0 & 2 & 5 \end{bmatrix}$.

a. Find the lengths of the sides of $\triangle PQR$.

- Is $\triangle PQR$ isosceles?
- Is $\triangle PQR$ a right triangle?

b. Find the coordinates of the midpoints of the sides of $\triangle PQR$.

3. Write equations for the lines satisfying the specified conditions:

a. Parallel to the line with equation $y = 0.25x - 4$ and passing through the point (4, 6)

b. Perpendicular to the line with equation $y = 0.25x - 4$ and y-intercept 3

c. Parallel to the line containing $A(2, 3)$ and $B(-4, 5)$ and containing $C(-2, -1)$

d. Perpendicular to the line containing $R(-4, -3)$ and $S(5, 2)$ and containing point S

3.2 Transformations

The rigid motions of *reflection, rotation,* and *translation* have coordinate representations, as do *size transformations*. By considering patterns that relate coordinates of image and pre-image points under a particular transformation, you can derive an algebraic representation of the transformation.

- Reflection across the x-axis: $(x, y) \rightarrow (x, -y)$
- Reflection across the y-axis: $(x, y) \rightarrow (-x, y)$
- Reflection across the line $y = x$: $(x, y) \rightarrow (y, x)$
- 90° counterclockwise rotation about the origin: $(x, y) \rightarrow (-y, x)$
- 180° rotation about the origin: $(x, y) \rightarrow (-x, -y)$
- Translation with components h and k: $(x, y) \rightarrow (x + h, y + k)$
- Size transformation with magnitude k and center at the origin: $(x, y) \rightarrow (kx, ky)$

EXAMPLE 1 ▶ Any rigid motion or size transformation of the plane that maps the origin onto itself always has a simple 2×2 matrix representation. The matrix representation can be found by finding the images of $A(1, 0)$ and $B(0, 1)$ and using the coordinates of the images as the columns of the transformation matrix.

For example, a 90° counterclockwise rotation about the origin maps $A(1, 0)$ onto $A'(0, 1)$ and $B(0, 1)$ onto $B'(-1, 0)$. Thus, the matrix representation for this rotation is $\begin{bmatrix} 0 & -1 \\ 1 & 0 \end{bmatrix}$.

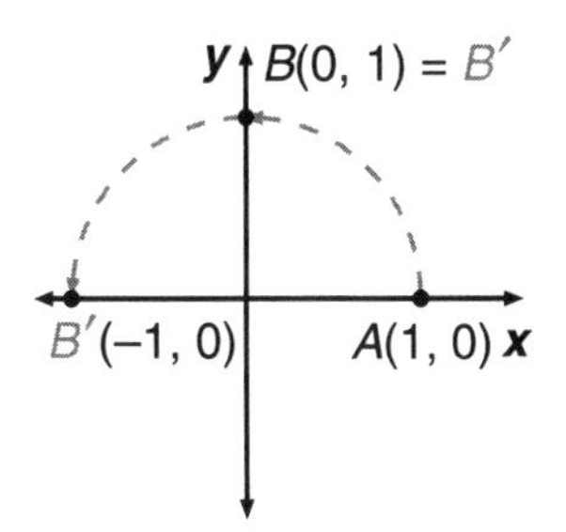

EXAMPLE 2 ▶ Some transformations of a polygon can be accomplished by matrix operations. Suppose $\triangle ABC$ has vertices $A(1, 2)$, $B(-3, 4)$, and $C(0, -3)$. Its image under a 90° counterclockwise rotation about the origin is found by multiplying the transformation matrix by the polygon matrix.

$$\begin{bmatrix} 0 & -1 \\ 1 & 0 \end{bmatrix} \begin{bmatrix} 1 & -3 & 0 \\ 2 & 4 & -3 \end{bmatrix} = \begin{bmatrix} -2 & -4 & 3 \\ 1 & -3 & 0 \end{bmatrix}.$$

Thus, the image triangle has vertices $A'(-2, 1)$, $B'(-4, -3)$, and $C'(3, 0)$. $\triangle A'B'C'$ is *congruent* to $\triangle ABC$. The two triangles have the same shape and are the same size.

EXAMPLE 3 ▶ *Similarity*—same shape, different size—is a central idea in many geometric applications. The simplest similarity transformation is the size transformation. Size transformations centered at the origin are represented by matrices of the form $\begin{bmatrix} k & 0 \\ 0 & k \end{bmatrix}$. When $k > 1$, the transformation is an *enlargement;* when $0 < k < 1$, the transformation is a *reduction*, and when $k = 1$, the transformation is the identity (it leaves all points in their original locations). The *magnitude* or *scale factor* of the size transformation is k. For example, when $k = 2.5$, the enlargement of $\triangle ABC$ in Example 2 is $\triangle A'B'C'$ with vertices $A'(2.5, 5)$, $B'(-7.5, 10)$, and $C'(0, -7.5)$.

$$\begin{bmatrix} 2.5 & 0 \\ 0 & 2.5 \end{bmatrix} \begin{bmatrix} 1 & -3 & 0 \\ 2 & 4 & -3 \end{bmatrix} = \begin{bmatrix} 2.5 & -7.5 & 0 \\ 5 & 10 & -7.5 \end{bmatrix}$$

EXAMPLE 4 ▶ An enlargement (or reduction) affects sides and angles of polygons in predictable ways. For the size change in Example 3, the image triangle has vertices $A'(2.5, 5)$, $B'(-7.5, 10)$, and $C'(0, -7.5)$. Since $AB = \sqrt{4^2 + (-2)^2} = \sqrt{20} = 2\sqrt{5}$ and $A'B' = \sqrt{10^2 + 5^2} = \sqrt{125} = 5\sqrt{5}$, $\frac{A'B'}{AB} = 2.5$. You can similarly show that $\frac{B'C'}{BC} = \frac{A'C'}{AC} = 2.5$, the scale factor for the transformation. Since the length of each side of $\triangle A'B'C'$ is 2.5 times the length of the corresponding side of $\triangle ABC$, the two triangles are *similar*. Thus, the corresponding angles are congruent. Note also that corresponding sides of $\triangle ABC$ and $\triangle A'B'C'$ are parallel, as you can show by computing the slopes.

Check Your Understanding 3.2

Complete the following problems to check your understanding of coordinates and matrix representations of transformations.

1. Write the matrix representation for each transformation.

a. Reflection across the x-axis

b. Rotation of 180° about the origin

c. Reflection across the line $y = -x$

d. Reflection across the y-axis

e. Counterclockwise rotation of 270° about the origin

f. Reduction by a factor of 0.4 with center at the origin

2. Show that the line through $A(2, 5)$ and $B(8, -3)$ and its image under a size transformation with magnitude 4.5 and center at the origin are parallel.

3.3 Trigonometry

In any two similar triangles, the ratio of corresponding sides is constant and is equal to the scale factor. In the case of similar right triangles, special *trigonometric ratios* can be defined. These ratios can be used to obtain measures of distances and angles even when the distances and angles cannot be measured directly.

$$\sin A = \frac{a}{c}$$

$$\cos A = \frac{b}{c}$$

$$\tan A = \frac{a}{b}$$

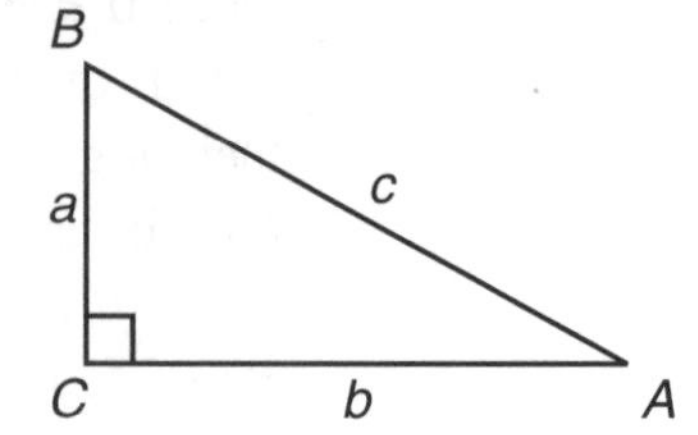

EXAMPLE 1 ▶ In right $\triangle ABC$, $m\angle B = 56°$ and $c = 12.1$. To find the length of $\overline{BC}$, use a trigonometric ratio that involves $m\angle B$, c, and a.

$$\cos 56° = \frac{a}{12.1}$$

So, $a = 12.1\cos 56°$

$\approx 12.1(0.5591929035)$

≈ 6.8

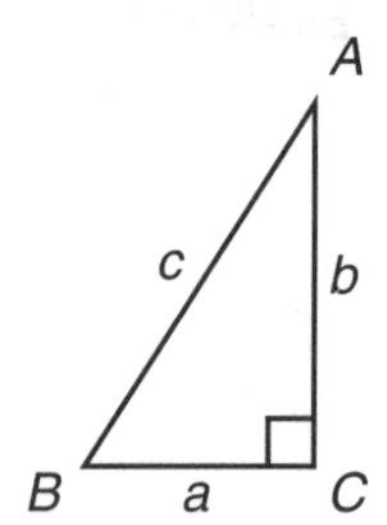

EXAMPLE 2 ▶ A tree casts a shadow of 40 feet when the *angle of elevation* to the sun is 35°. To find the height h of the tree, you can use the fact that $\tan 35° = \frac{h}{40}$. So, $h = 40 \tan 35° \approx 28.0$ feet.

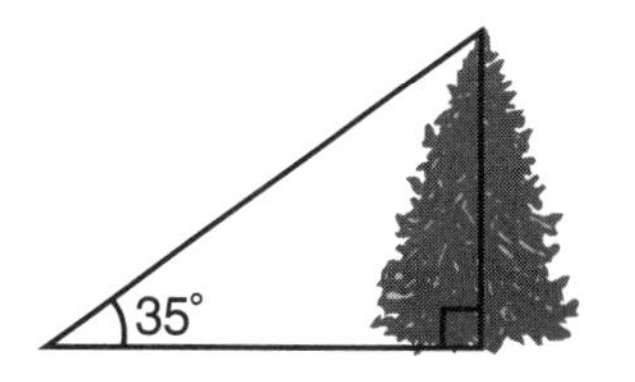

EXAMPLE 3 ▶ The trigonometric ratios of acute angles with measures 30°, 45°, and 60° can be computed exactly. For example, in the following diagram, $\triangle ABC$ is an equilateral triangle. It follows that $m\angle DBA$ is 30° and $AD = \frac{1}{2}AC$ since $\triangle ABD$ is half of the equilateral triangle. Thus, letting $AD = 1$, the length AB is 2 and BD is $\sqrt{4-1} = \sqrt{3}$.

So, $\sin 30° = \dfrac{1}{2}$,

$\cos 30° = \dfrac{\sqrt{3}}{2}$,

$\tan 30° = \sqrt{\dfrac{1}{3}}$ or $\dfrac{1}{\sqrt{3}} \cdot \dfrac{\sqrt{3}}{\sqrt{3}} = \dfrac{\sqrt{3}}{3}$.

Verify that $\sin 60° = \dfrac{\sqrt{3}}{2}$, $\cos 60° = \dfrac{1}{2}$, and $\tan 60° = \sqrt{3}$.

In the diagram below, if the measure of $\angle B$ is 45°, then

$AC = CB$. Let $AC = 1$. Then $AB = \sqrt{2}$. So, $\sin 45°$ is $\dfrac{1}{\sqrt{2}}$ or $\dfrac{\sqrt{2}}{2}$, $\cos 45° = \dfrac{\sqrt{2}}{2}$, and $\tan 45° = \dfrac{1}{1} = 1$.

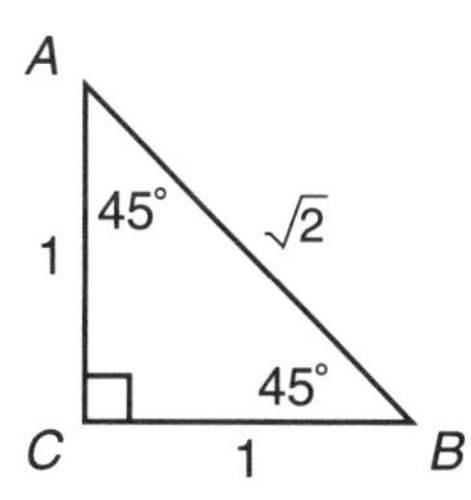

EXAMPLE 4 ▶ Angles can be measured in either *degrees or radians*. A degree is $\frac{1}{360}$ of a complete turn about a point. A radian is $\frac{1}{2\pi}$ of a complete rotation. So, 1 revolution = 360° = 2π radians. To convert between degrees and radians, solve this proportion:

$$\frac{\text{measure of angle in radians}}{2\pi} = \frac{\text{measure of angle in degrees.}}{360}$$

For example, to convert 30° to radians, solve:

$$\frac{\text{measure of angle in radians}}{2\pi} = \frac{30}{360}.$$

So, $30° = \frac{\pi}{6}$ radians.

EXAMPLE 5 ▶ Some of the important applications of radian measure involve *linear* and *angular velocity*.

If a point P moves in uniform circular motion about a point O as, for example, a point on the edge of a pulley, then the linear velocity of P is $V = \frac{s}{t}$ where s is the length of the arc traversed and t is the time required.

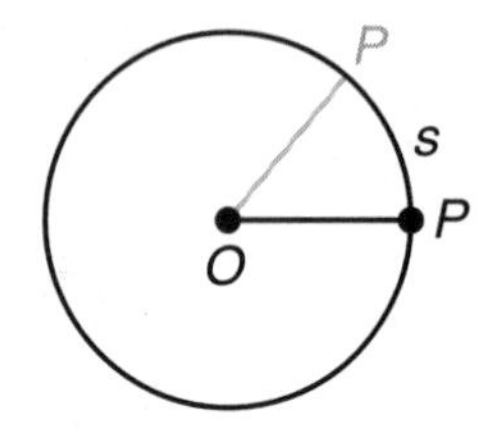

a. The linear velocity in meters per second of a horse 6 meters from the center of a merry-go-round that is turning at a rate of 12 revolutions per minute (rpm) is $(6\text{m})(2\pi)(12)\frac{\text{m}}{\text{min}} = \frac{144\pi}{60}\frac{\text{m}}{\text{sec}} = \frac{12\pi}{5}\frac{\text{m}}{\text{sec}}$.

The angular velocity of $\overline{OP}$ is $w = \frac{\theta}{t}$ where θ is the measure of the angle generated by $\overline{OP}$ and t is the time required.

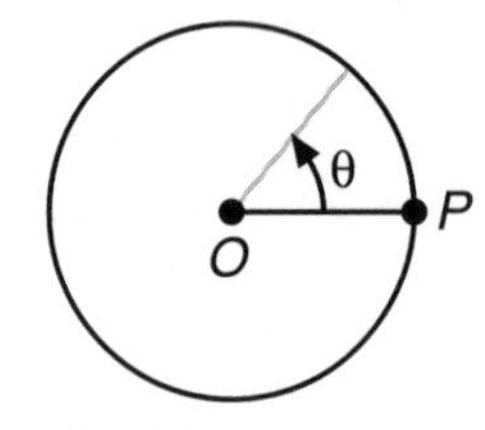

b. The angular velocity in radians per second of the merry-go-round is 12 rpm $= 12(2\pi)\frac{\text{rad}}{\text{min}} = 24\pi\frac{\text{rad}}{\text{min}} = 24\pi\frac{\text{rad}}{\text{min}} \cdot \frac{1 \text{ min}}{60 \text{ sec}} = \frac{24\pi}{60}\frac{\text{rad}}{\text{sec}} = \frac{2\pi}{5}\frac{\text{rad}}{\text{sec}}$.

EXAMPLE 6 ▶ The definitions of sine and cosine generalize to angles with measures greater than or equal to $90°\left(\frac{\pi}{2}\right)$ and to angles of negative measure. Think of point $A(1, 0)$ on a *unit circle* rotating about the origin. If the rotation is counterclockwise, as in the figures on the next page, the angle of rotation is positive. If the rotation is clockwise, the angle is negative. In this context, the *sine function* measures the vertical position of the point.

For example, suppose point A rotates counterclockwise $\frac{\pi}{4}(45°)$.

Then, $\sin \frac{\pi}{4} = \frac{\frac{\sqrt{2}}{2}}{1} = \frac{\sqrt{2}}{2}$.

Note $\sin \frac{\pi}{2} = \frac{1}{1} = 1$.

In general, $\sin q = \frac{y}{1} = y$.

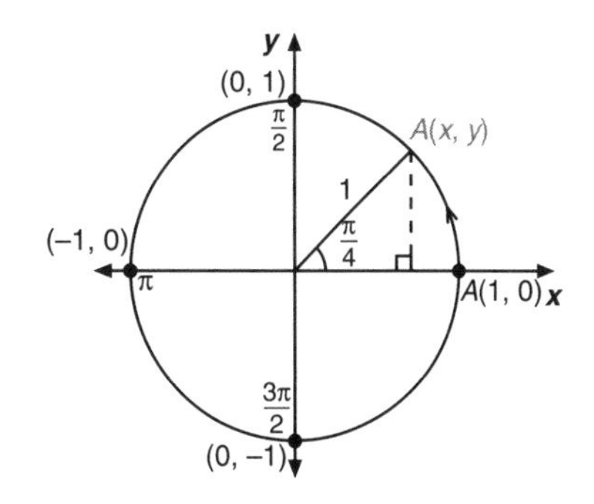

Similarly, the *cosine function* measures the horizontal position of the point. For example, suppose point A rotates counterclockwise $\frac{5\pi}{4}$.

Then, $\cos \frac{5\pi}{4} = \frac{-\frac{\sqrt{2}}{2}}{1} = \frac{-\sqrt{2}}{2}$.

Note $\cos \frac{3\pi}{2} = \frac{0}{-1} = 0$.

In general, $\cos \theta = \frac{x}{1} = x$.

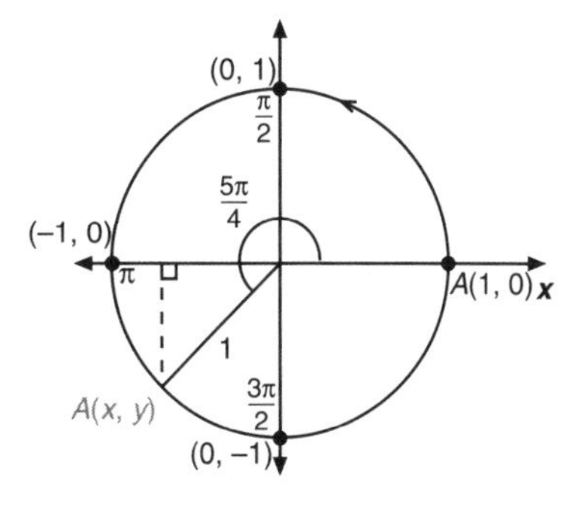

Verify the other entries in the following table.

θ	0	$\frac{\pi}{4}$	$\frac{\pi}{2}$	$\frac{3\pi}{4}$	π	$\frac{5\pi}{4}$	$\frac{3\pi}{2}$	$\frac{7\pi}{4}$	2π
sin θ	0	$\frac{\sqrt{2}}{2}$	1	$\frac{\sqrt{2}}{2}$	0	$-\frac{\sqrt{2}}{2}$	-1	$-\frac{\sqrt{2}}{2}$	0
cos θ	1	$\frac{\sqrt{2}}{2}$	0	$-\frac{\sqrt{2}}{2}$	-1	$-\frac{\sqrt{2}}{2}$	0	$\frac{\sqrt{2}}{2}$	1

The trigonometric functions sine and cosine are *periodic functions* with period 2π as shown in the graphs below.

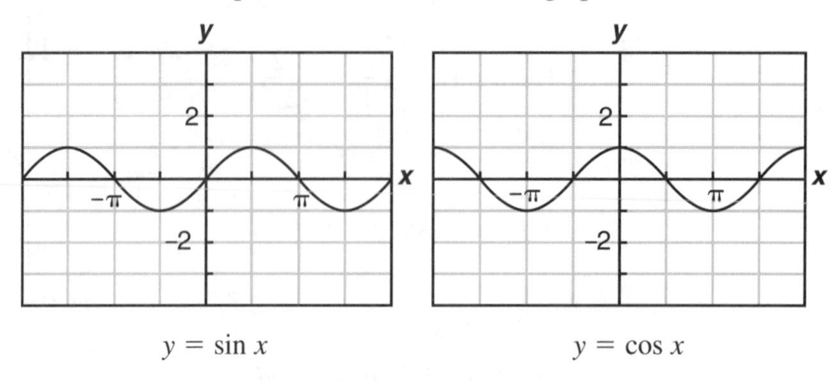

$y = \sin x$ $y = \cos x$

EXAMPLE 7 ▶ Variations of the sine and cosine functions have predictable graphs. Note that the graph of $y = 2\sin x$ has the same shape as the graph of $y = \sin x$; but is stretched vertically. The *amplitude* of a periodic function with maximum value M and minimum value m is $\frac{1}{2}\,|M - m|$. The amplitude of $y = \sin x$ is 1. The amplitude of $y = 2\sin x$ is 2.

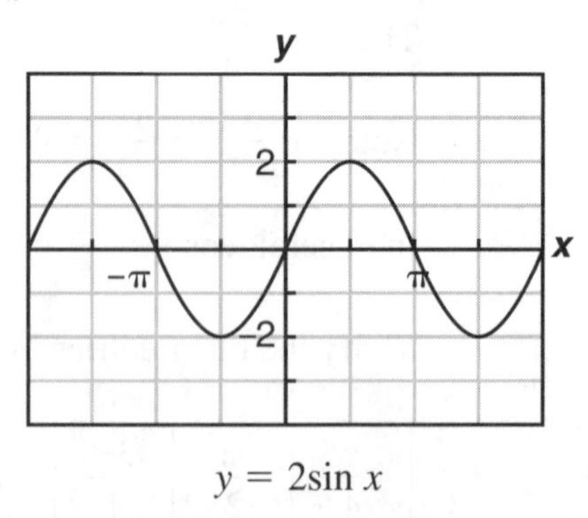

$y = 2\sin x$

Check Your Understanding 3.3

Check your understanding of trigonometric ratios and functions and radian measure by completing the following tasks.

1. A building casts a shadow of 100 feet when the sun is at an angle of elevation of 60°. Use this information to find the height of the building.

2. Consider right triangle PQR shown at the right.

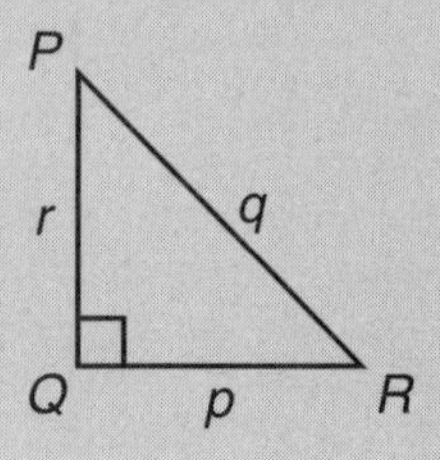

a. Express q in terms of p and r.

b. Find the values of the three trigonometric ratios for $\angle R$ when p is 3 and q is 8.

c. Suppose $\sin P = \frac{7}{25}$. Find three triples of values that p, q, and r could take on.

3. Complete the following table, giving exact values where possible.

Measure of angle	30°	45°	75°
Sine of angle			
Cosine of angle			
Tangent of angle			

4. Write the equivalent radian measure for each degree measure.

a. 30° **b.** 45° **c.** 60°

d. 90° **e.** 180° **f.** 270°

5. Write the equivalent degree measure for each angle measure given in radians.

a. $\frac{\pi}{6}$ **b.** $\frac{\pi}{3}$ **c.** $\frac{\pi}{4}$

d. $\frac{3\pi}{4}$ **e.** $\frac{2\pi}{3}$ **f.** $\frac{11\pi}{6}$

6. A ceiling fan has a diameter of 48 inches. If the fan is turning at 80 rpm, find:

a. Its angular velocity in radians per second.

b. The linear velocity in inches per second of a point on the tip of a blade.

7. Think about the relationship between the form of trigonometric function rules and the shapes of their graphs.

a. How is the graph of $y = 3\sin x$ related to the graph of $y = \sin x$?

b. How is the graph of $y = -3\cos x$ related to the graph of $y = \cos x$?

4 Discrete Mathematics

Nearly every business, industry, or branch of government has to plan work on complex problems. They need to make decisions that will solve problems that often encompass many variables and at the same time will efficiently use materials, money, time, etc. Matrix models and vertex-edge graph models are often helpful in solving such problems.

In earlier study of these models, you learned how to use discrete mathematical methods to answer questions like these:

- After a snowstorm, how can a county decide which roads to plow so that it is possible to travel from every town to every other town on plowed roads, and the total number of miles plowed is minimum?
- What route should a salesperson travel to visit each client once and make the travel distance or cost minimal?
- What is an efficient way of tracking inventories of products?
- How can pollution be tracked through an ecosystem?

4.1 Matrix Models

In Course 1, you used adjacency matrices to represent vertex-edge graphs. In the algebra and geometry review sections of this book, you revisited how matrices can be used to solve systems of linear equations and to represent polygons and geometric transformations. A *matrix* is a rectangular array of numbers. The matrix $A = \begin{bmatrix} 2 & 4 & 5 \\ -3 & 1 & 7 \end{bmatrix}$ has *dimension* 2 by 3 since it has two rows and three columns. A *square matrix* has the same number of rows and columns. The *main diagonal* of a square matrix is the diagonal line of entries running from the top left corner to the bottom right corner.

EXAMPLE 1 *Under The Willow* is a toy company in Seattle that makes stuffed toys, including rabbits, frogs, and willow elves. The owner designs the toys, and then they are cut out, sewn, and stuffed by independent contractors. Each contractor agrees to make the number of stuffed toys shown in the following matrix for the months of September and October.

Number of Toys to Make

$$\begin{array}{r} \\ \text{Rabbits} \\ \text{Frogs} \\ \text{Elves} \end{array} \begin{array}{c} \begin{array}{cc} \text{Sept} & \text{Oct} \end{array} \\ \begin{bmatrix} 10 & 20 \\ 25 & 30 \\ 10 & 30 \end{bmatrix} \end{array}$$

Two of the contractors, Elise and Harvey, know from experience how many minutes it takes them to make each type of toy. The times are shown in the matrix below.

Time per Toy (in minutes)

$$\begin{array}{r} \\ \text{Elise} \\ \text{Harvey} \end{array} \begin{array}{c} \begin{array}{ccc} \text{Rabbit} & \text{Frog} & \text{Elf} \end{array} \\ \begin{bmatrix} 55 & 60 & 90 \\ 80 & 50 & 100 \end{bmatrix} \end{array}$$

a. Matrix multiplication can be used to find a matrix that shows the total number of minutes each of the two contractors will need in order to fulfill their contracts for each of the two months, as shown below.

$$\begin{bmatrix} 55 & 60 & 90 \\ 80 & 50 & 100 \end{bmatrix} \times \begin{bmatrix} 10 & 20 \\ 25 & 30 \\ 10 & 30 \end{bmatrix} = \begin{bmatrix} 2{,}950 & 5{,}600 \\ 3{,}050 & 6{,}100 \end{bmatrix}$$

Thinking of the labels for the product matrix helps you interpret the entries. The row labels, Elise and Harvey, come from the first matrix in the indicated product. The column labels, Sept and Oct, come from the second factor. So, for example, Harvey will need 3,050 minutes to fulfill September orders and 6,100 minutes to fill October orders.

b. Minute totals can be converted to hours using scalar multiplication as shown below.

$$\frac{1}{60} \times \begin{bmatrix} 2{,}950 & 5{,}600 \\ 3{,}050 & 6{,}100 \end{bmatrix} = \begin{bmatrix} 49.2 & 93.3 \\ 50.8 & 101.7 \end{bmatrix}$$

EXAMPLE 2 ▶ Similar to operations with numbers, matrices can be combined using the operations of addition, subtraction, and matrix multiplication (as seen above). Consider the following matrices.

$$A = \begin{bmatrix} 4 & 3 \\ -2 & 2 \end{bmatrix} \qquad B = \begin{bmatrix} 1 \\ 6 \end{bmatrix} \qquad C = \begin{bmatrix} 3 & 9 \\ -5 & -8 \end{bmatrix}$$

a. In order to add or subtract two matrices, they must have the same dimension. To add (or subtract) two matrices, you add (or subtract) corresponding entries.

$$A + C = \begin{bmatrix} 7 & 12 \\ -7 & -6 \end{bmatrix} \qquad A - C = \begin{bmatrix} 1 & -6 \\ 3 & 10 \end{bmatrix}$$

b. In order to multiply two matrices, the number of columns in the first matrix must equal the number of rows in the second. Matrix multiplication is *not* commutative. The process of matrix multiplication is illustrated below.

$$AB = \begin{bmatrix} 4 & 3 \\ -2 & 2 \end{bmatrix} \begin{bmatrix} 1 \\ 6 \end{bmatrix} = \begin{bmatrix} 4 \cdot 1 + 3 \cdot 6 \\ -2 \cdot 1 + 2 \cdot 6 \end{bmatrix} = \begin{bmatrix} 22 \\ 10 \end{bmatrix}$$

c. It is also possible to multiply a matrix can be multiplied by a number k. This is done by multiplying each entry in the matrix by k. This is called *scalar multiplication.*

$$3A = \begin{bmatrix} 12 & 9 \\ -6 & 6 \end{bmatrix}$$

EXAMPLE 3 ▶ Vertex-edge graphs are often used to model transportation networks such as airline routes. The network shown below and its adjacency matrix indicate airline routes between various cities.

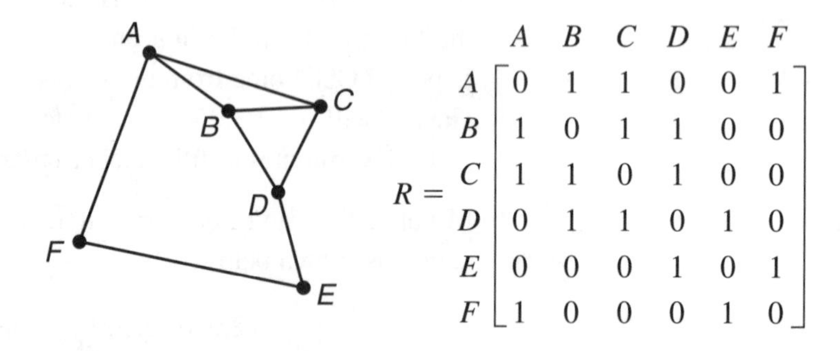

$$R = \begin{array}{c} \\ A \\ B \\ C \\ D \\ E \\ F \end{array} \begin{array}{c} \begin{matrix} A & B & C & D & E & F \end{matrix} \\ \begin{bmatrix} 0 & 1 & 1 & 0 & 0 & 1 \\ 1 & 0 & 1 & 1 & 0 & 0 \\ 1 & 1 & 0 & 1 & 0 & 0 \\ 0 & 1 & 1 & 0 & 1 & 0 \\ 0 & 0 & 0 & 1 & 0 & 1 \\ 1 & 0 & 0 & 0 & 1 & 0 \end{bmatrix} \end{array}$$

Squaring the adjacency matrix ($R^2 = R \times R$) shows the number of two-stage routes connecting various cities. For example, there are two two-stage paths from city A to city D.

$$R^2 = \begin{bmatrix} 3 & 1 & 1 & 2 & 1 & 0 \\ 1 & 3 & 2 & 1 & 1 & 1 \\ 1 & 2 & 3 & 1 & 1 & 1 \\ 2 & 1 & 1 & 3 & 0 & 1 \\ 1 & 1 & 1 & 0 & 2 & 0 \\ 0 & 1 & 1 & 1 & 0 & 2 \end{bmatrix}$$

Similarly, $R^3 = R \times R \times R$ shows the number of three-stage routes connecting cities.

Check Your Understanding 4.1

Check your understanding of matrices and operations on them.

1. Using the given matrices, find the indicated sum, difference, and products.

$$A = \begin{bmatrix} 1 & 6 & 2 \\ 3 & 5 & -1 \end{bmatrix} \qquad B = \begin{bmatrix} 0 & -5 & 1 \\ 3 & 2 & 4 \end{bmatrix} \qquad C = \begin{bmatrix} 2 & -4 \\ 1 & 1 \end{bmatrix}$$

a. $A + B$ **b.** $B - A$ **c.** $-2A$ **d.** CA

2. During a recent inventory, Just Jeans found that they had 35 pairs of Levis in stock in the following waist sizes.

Waist Size	Number of Pairs
28	6
30	14
32	9
34	6

For other brands carried by the store, the number of pairs in comparable sizes ordered from smallest to largest is shown in the chart below.

Brand	Number of Pairs
Tommy Hilfiger	0, 4, 0, 5
Polo	1, 2, 3, 8
Wrangler	6, 2, 2, 2
JNCO	3, 0, 0, 4
Calvin Klein	7, 4, 1, 5

a. Represent the current Just Jeans inventory using a matrix.

- How many pairs of jeans in stock are size 32?
- Of which brands are there the fewest jeans?

b. Sales over the last year indicate that the store sells twice as many Tommy Hilfiger and Calvin Klein jeans as the other four brands. Sizes 30 and 32 sell twice as fast as sizes 28 and 34. Suppose the stock is filled in so that Just Jeans now has inventory of 12 pairs of jeans in each of the most popular sizes of the best-selling brands. Represent the restocked inventory with a matrix.

c. In the following week, the store sells the following:

$$S = \begin{array}{c} \\ 28'' \\ 30'' \\ 32'' \\ 34'' \end{array} \begin{array}{c} \begin{array}{cccccc} L & TH & P & W & J & CK \end{array} \\ \begin{bmatrix} 2 & 3 & 0 & 2 & 2 & 4 \\ 4 & 8 & 2 & 2 & 0 & 10 \\ 5 & 10 & 2 & 2 & 0 & 9 \\ 2 & 5 & 1 & 1 & 1 & 3 \end{bmatrix} \end{array}$$

The profit per pair of jeans is $16.00 on Levi, $20 on Tommy Hilfiger, $18.50 on Polo, $15 on Wrangler, $12.00 on JNCO, and $16 on Calvin Klein.

- Use matrix operations to determine how much profit the store made that week on the largest size sold. How much profit was made on the smallest size sold?
- Determine the total profit on jeans sales for the week.

3. Construct the route (adjacency) matrix for the network below.

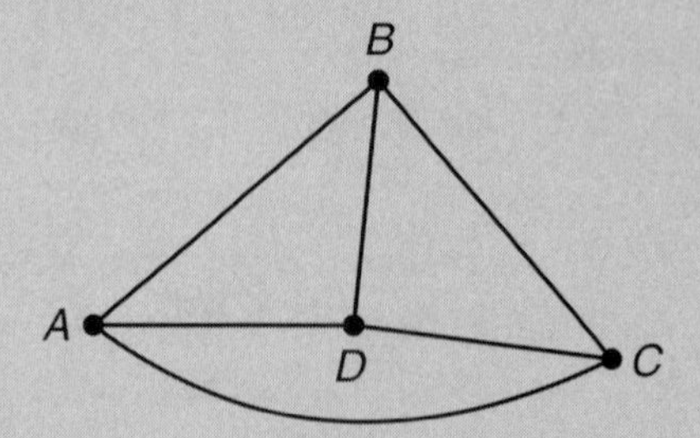

a. Find the matrix that describes the two-stage paths between vertices, and then give a vertex sequence for the various routes from D to C.

b. How many three-stage paths start and finish at the same point? How many three-stage paths start and finish at different points?

4.2 Network Optimization

Finding a solution to a problem modeled by a vertex-edge graph may involve optimizing the network in some way.

EXAMPLE 1 ▶ In a nature center, there are seven stations to view unique plant growth. The seven stations are labeled *A–G* and are connected by gravel paths that have the following lengths (in meters).

	A	***B***	***C***	***D***	***E***	***F***	***G***
A	0	20	40	50	50	–	–
B	20	0	30	30	70	70	–
C	40	30	0	40	50	60	50
D	50	30	40	0	20	20	30
E	50	70	50	20	0	20	30
F	–	70	60	20	20	0	50
G	–	–	50	30	30	50	0

The plan is to pave some paths with asphalt so that each station is wheelchair accessible. Because of a limited budget, the nature center directors want the number of meters of pavement to be as small as possible. The nature center trails can be represented by a vertex-edge graph where vertices are stations and edges represent paths connecting stations. Since there are lengths associated with the edges, the graph is a *weighted graph.*

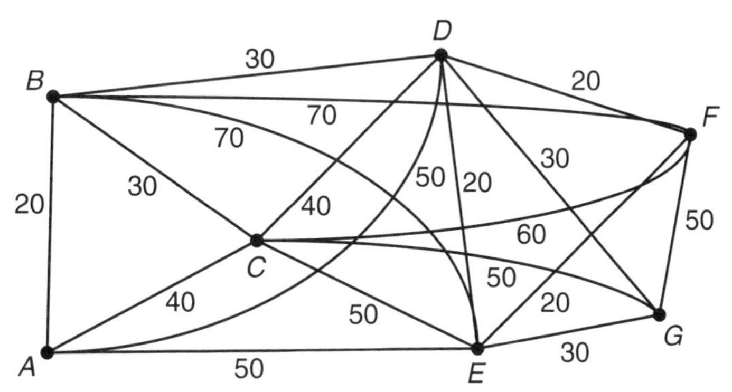

A *minimal spanning tree* is a *tree* (a connected graph with no circuits) that spans all the vertices and has a minimum total length. Finding a minimal spanning tree will solve the nature center problem. You can find a minimal spanning tree using the *best-edge*

algorithm: Choose a shortest edge that does not complete a circuit, mark it, and then choose again a shortest edge that makes no circuit, until no additional choices can be made. The result is a minimal spanning tree as shown.

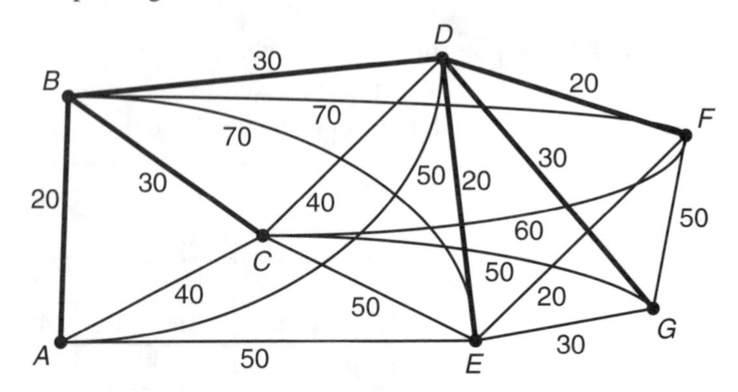

You could find other minimal spanning trees by making different choices of the 20-meter or 30-meter paths to pave. Any minimal spanning tree of this graph has a length of 150 meters.

EXAMPLE 2 ▶ Consider the vertex-edge graph below. Notice that there is no *Euler circuit* (a circuit that traverses each edge exactly once) for the graph, since there are vertices of odd degree. However, the graph does have a *Hamiltonian circuit* as shown (that is, a circuit that visits each vertex once and only once, except the first vertex). There are no known necessary and sufficient conditions for a Hamiltonian circuit.

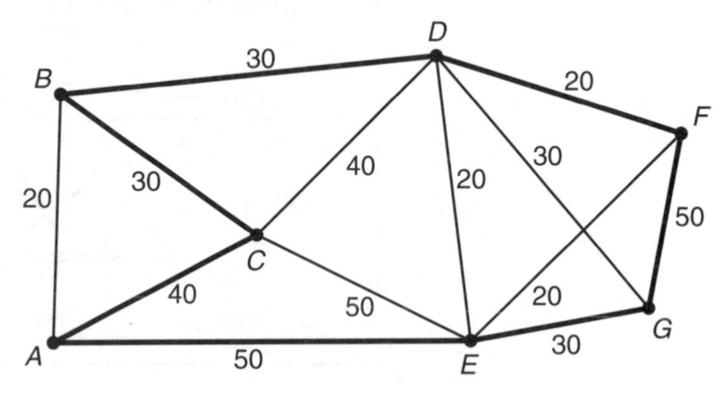

EXAMPLE 3 ▶ The length of the Hamiltonian circuit above is 250. Finding a minimum-length Hamiltonian circuit is called the *Traveling Salesperson Problem*. One solution to the traveling salesperson problem related to the network above is shown on the graph

below. The circuit has a length of 240. Often the shortest Hamiltonian circuit is not unique.

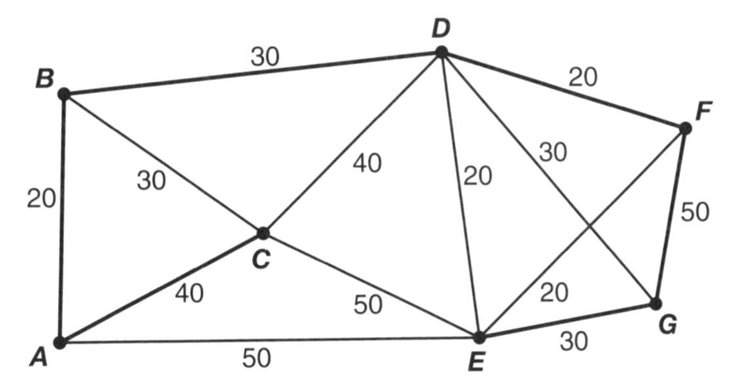

Check Your Understanding 4.2

Complete the following problems to check your understanding of spanning trees and circuits.

1. For the graph of Example 2 on page 42, find a Hamiltonian circuit that contains the edge *E–F* and has a length of 240.

2. Find a minimal spanning tree for the graph shown below.

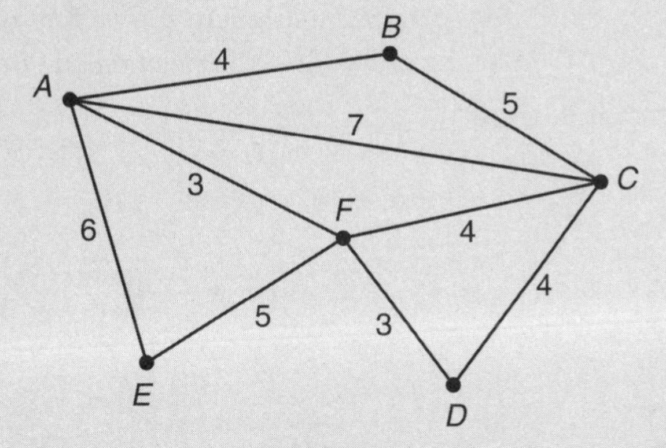

3. Suppose cities are located at each vertex of the graph in Problem 2. A transport driver needs to visit each city in the smallest number of miles and return to home base at *A*. Find a route for the transport driver.

4. A road grader needs to travel each road in the graph of Problem 2. Can this be done without traveling any road more than once? If so, show a circuit to follow.

Maintaining Concepts and Skills

The following sets of exercises give you an opportunity to review the mathematical ideas and skills that you acquired in Course 2 units and early Course 3 units. They include questions that require a knowledge of:

- algebra and functions
- geometry and trigonometry
- statistics and probability
- discrete mathematics

Some problems combine ideas from two or more of those strands of mathematics. In each case, you will need to determine the appropriate ideas and techniques to apply.

If you need a refresher on some particular topic, look back at the reference material and examples in the first section of this guide. However, it is best to make a good effort at completing an exercise before looking for help in the reference material or in the answers that are given at the back of the book. Since practice of any skill is most effective when distributed in modest amounts throughout the school year, the practice exercises have been arranged in sets of ten items so you could do about one set every other week throughout the course as companion work to your study of new topics in *CMIC* Course 3. Beginning with Exercise Set 11, exercise sets include some questions over material you studied in early Course 3 units. You should complete those exercises during the second half of the course.

Exercise Set 1

1. A cellular communications company charges \$20 per month and \$0.075 per minute of use.

a. What equation shows how to calculate the monthly bill y as a function of the number of minutes of use x?

b. What equation using *NOW* and *NEXT* shows how the monthly bill grows as each additional minute of calling time is used?

c. Find the time the phone was used if a monthly bill was \$38.15.

d. Find the cost for a month if you use the phone for 372 minutes.

e. Find the number of calling minutes possible if a monthly bill is to remain below \$55.

f. A second service has a monthly charge of \$10 and charges \$0.10 per minute of use. For what monthly calling times is the second service the better choice?

2. Solve each linear system by reasoning with the symbols themselves. Check each result.

a. $2x + y = 7$
$3x - y = 3$

b. $y = 3x + 6$
$2y = -3x + 3$

c. $2x + 3n = 9$
$5x - 3n = 5$

3. $\triangle ABC$ has vertices $A(1, 2)$, $B(4, -2)$, and $C(-2, -1)$.

a. Find the length of each side.

b. Is $\triangle ABC$ a right triangle? Support your answer with appropriate reasoning.

c. Find the coordinates of the midpoints of each side of $\triangle ABC$.

d. Find the equation of the line containing each side.

4. Solve each inequality for x.

a. $-3x + 7 \geq 1$

b. $x^2 - 4 \geq 0$

c. $2(x + 1) \leq 3x + 9$

d. $2^x \geq 1$

e. $4x^2 < 24$

5. Rewrite each expression in an equivalent and simplified form using only positive exponents. Assume all variables are nonzero.

a. $5^3 \cdot 5^{-2}$

b. $q^{-4} \cdot q^2$

c. $\frac{2^{-4}}{2^5}$

d. $4^0 \cdot 4^{-2}$

e. $(xy^2)^4$

f. $(x^{-2})^3$

Exercise Set 1

6. A flagpole in front of Central High School stands 80 feet tall.

a. Find the length of its shadow when the angle of elevation from the end of the shadow to the sun is 20°.

b. Suppose a support cable is to be attached 40 feet up the flagpole and makes an angle of 30° with the ground. How far from the base of the flagpole would the cable be attached to the ground?

c. How long is the cable in Part b? Find this length in two ways.

7. In a new theater design, the number of seats in each row is 16 fewer than the number of rows.

a. Write an equation giving the number of seats as a function of the number of rows.

b. Estimate the number of rows if there are 1,161 seats.

c. Find the number of seats in each row and the number of rows in a planned 2,145 seat auditorium.

8. Consider the quadratic function defined by $y = x^2 - 4x + 3$.

a. Find the x- and y-intercepts of its graph.

b. Find the equation of the symmetry line of its graph.

c. Which way does its graph open? Explain your answer.

d. Sketch the graph of $y = x^2 - 4x + 3$.

9. Write a matrix representation for each transformation:

a. Reflection across the y-axis

b. Size transformation with magnitude 2 and center at the origin

c. Reflection across the line $y = x$

d. Counterclockwise rotation of 90° about the origin

10. Suppose you roll two dice until you get a sum of 6.

a. What is the probability you get a sum of 6 for the first time on the fourth roll?

b. How many times do you expect to have to roll the dice until you get a sum of 6?

Exercise Set 2

1. Suppose, as part of an agreement with her father to do some work for him during the summer, Ellen will receive 1¢ for the first day of work, but every day after that her pay will double.

   a. What equation shows how to calculate Ellen's daily pay p as a function of the number of days n she has worked?

   b. What equation using *NOW* and *NEXT* shows how Ellen's pay grows as each additional day of work passes?

   c. If Ellen's pay for a day is $10.24, how many days has she worked?

   d. Find Ellen's daily pay if she has worked 20 days.

   e. For how many days will Ellen earn less than $20 per day?

2. Metro Cab company charges a base price of $2.50 plus 20¢ per mile. A competitor, National Cab company, charges a base price of $1.50 plus 30¢ per mile.

   a. Write equations modeling the costs for both cab companies.

   b. If you need to travel 5 miles, which cab company is the least expensive?

   c. If you need to travel 15 miles, which cab company is the least expensive?

   d. For what trip length are the costs the same for the two cab companies?

3. Triangle ABC has vertices $A(2, -4)$, $B(-6, 5)$, and $C(1, 4)$. Find the coordinates of the vertices of the image $\triangle A'B'C'$ when $\triangle ABC$ is reflected

   a. Across the x-axis. **b.** Across the y-axis. **c.** Across the line $y = x$.

4. Solve each quadratic equation by reasoning with the symbols themselves.

   a. $x^2 - 48 = 0$ **b.** $4x^2 + 7 = 19$

   c. $2x^2 + 8 = 30$ **d.** $\frac{1}{3}x^2 - 4 = -1$

5. Consider the matrices:

$$A = \begin{bmatrix} 6 & 9 \\ 2 & 0 \end{bmatrix} \qquad B = \begin{bmatrix} -4 & 7 \\ -12 & 3 \end{bmatrix}$$

   a. Find $A + B$. **b.** Evaluate $A - B$. **c.** Calculate AB.

6. $\triangle ABC$ is a right triangle with vertices $A(1, 2)$, $B(5, 2)$, and $C(5, x)$. Point C is in the first quadrant.

   a. If the length AC is 5, find the length of each side.

   b. Find the equation of the line containing each side.

Exercise Set 2

7. Kim is flying his box kite and has let out 400 feet of string. His end of the string is 4 feet off the ground.

a. If the angle of elevation to the kite is 50°, approximately how high off the ground is the kite?

b. Kim pulls in some string and notices that the kite is right above his house which is 210 feet away. If the angle of elevation to the kite is still 50°, approximately how high is the kite?

8. Consider the graph shown at the right.

a. Find a minimal spanning tree.

b. What is the length of the minimal spanning tree?

c. Find a Hamiltonian circuit for the graph.

d. What is the length of the shortest Hamiltonian circuit?

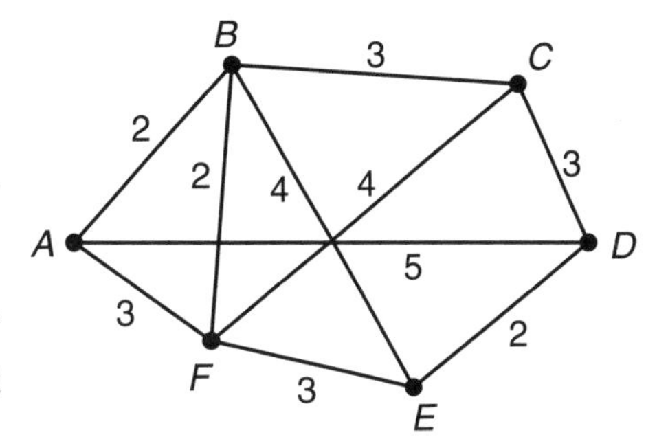

9. A rectangle is 6 feet longer than it is wide.

a. Write an equation giving the area of the rectangle.

b. Estimate the width, in feet, if the area is 111 square feet.

c. If the area is 216 square feet, what is the perimeter of the rectangle?

10. The data below give the number of minutes each of nine children watched a video before wandering off to other activities.

Age in Years, *x*	2	3	3.2	5	4.1	3.7	2.2	4.7	5.6
Time in Minutes, *y*	1	2.5	2	7	6	6.5	1.5	5.3	7.3

a. Is this set of data reasonably modeled by a line? Explain.

b. Find the equation of the regression line for the data.

c. What is the slope of the regression line? Explain the meaning of the slope in the context of the data.

d. Find the residual for $x = 2$.

Exercise Set 3

1. Rewrite each of the following expressions in the form $ax + by = c$ where a, b, and c are integers.

a. $6x - 3 = y$ **b.** $2(y + x) + 5 = 0$ **c.** $2x + 4y = \frac{1}{3}$

d. $\frac{1}{3}y + 5 = 2x$ **e.** $3y = \frac{1}{5}x - 2$ **f.** $\frac{1}{3}x + \frac{1}{5}y = 2$

2. Solve each linear system by reasoning with the symbols themselves. Check each result.

a. $y = x + 5$
$y = 2x + 4$

b. $3x + 4y = 10$
$x - y = 1$

c. $4x + 3y = 12$
$3y - 8x = -12$

3. Sketch graphs of each of the following equations.

a. $y = x^2$ **b.** $y = x^2 + 2$ **c.** $y = -x^2 + 2$

4. A local newspaper reported that for the annual community theater summer production, 500 tickets were sold for a total of \$5,100. Tickets were priced at \$15 for adults and \$7 for children. How many adult's tickets, and how many children's tickets were sold for the production?

5. Draw sample scatterplots that contain:

a. An influential point that increases the correlation coefficient.

b. An influential point that decreases the correlation coefficient.

6. Use calculator-produced tables or graphs to solve each equation.

a. $29 = x^2 + 5x + 5$ **b.** $3x^2 = 5x + 2$ **c.** $4x = x^2 + 4$

7. Sketch graphs of each of the following periodic functions for $-360° \leq x \leq 360°$.

a. $y = \cos x$ **b.** $y = 3\cos x$

c. $y = \sin x$ **d.** $y = -2\sin x$

Exercise Set 3

8. As an incentive to attend the winter pep rally, King High School is giving away a gift certificate to some lucky student in attendance. The name of the winner of the gift certificate will be chosen randomly from all students enrolled in the school, but you must be present at the rally to win. Names will be drawn with replacement until someone wins. Approximately 60% of the student body usually attends the pep rally.

a. Make a probability distribution table for the number of names drawn until a winner is selected.

b. Make a graph of the probability distribution.

c. What is the probability that the fifth name drawn will win the gift certificate?

d. Would it be a rare event to have to draw four names?

9. A cylindrical container has a radius of r cm and a height of 10 cm.

a. Complete the following table.

Radius (cm)	1	2	3	4	7	r
Volume (cm^3)						

b. Draw a graph of the volume as a function of the radius.

c. Describe the pattern of change in the graph of the (*radius*, *volume*) data.

d. What happens to the volume when the radius is tripled?

10. A Ferris wheel with a radius of 10 feet is rotating counterclockwise at 2 revolutions per minute. Jan gets on the Ferris wheel at its lowest point, which is 2 feet above the ground. Assuming the Ferris wheel doesn't stop, how high above the ground is Jan:

a. After 10 seconds? **b.** After 30 seconds? **c.** After 50 seconds?

Exercise Set 4

1. Rewrite each expression in the form $y = \ldots$ when $a = 2$, $b = 3$, $c = 4$, and $d = -1$.

a. $x = \frac{a}{2}(y + b)$

b. $x = \frac{b - ay}{1 - c}$

c. $ax + by = cx + d$

d. $x = a\frac{by}{c^2}$

e. $x = \frac{y}{c + d}$

f. $(a - b)x - (a + b)y + dc^2 = 0$

2. Solve the linear system of equations.

$$5.85x + 3.13y = 2$$

$$17.27x + 2.13y = 3.65$$

3. Write each expression in an equivalent and simplified form. Assume all variables are greater than zero.

a. $\sqrt{16}\sqrt{81}$

b. $\sqrt{8}\sqrt{12}$

c. $\left(\frac{25}{64}\right)^{\frac{1}{2}}$

d. $3\sqrt{22}\sqrt{169}$

e. $3\sqrt{21} \cdot 4\sqrt{33}$

f. $2\sqrt{16}$

g. $\sqrt{4x^6y^2}$

h. $6\sqrt{8r^9}$

4. Consider the periodic function $y = 2\cos x$, for $0° \leq x \leq 360°$.

a. Find the *x*- and *y-intercepts* of the graph of this function.

b. What is the maximum value of this function, and at what point(s) does it occur?

c. What is the minimum value of this function, and at what point(s) does it occur?

5. $\triangle ABC$ has vertices $A(1, 4)$, $B(3, 5)$, and $C(4, 5)$.

a. Find the length of each side.

b. Find the coordinates of the midpoints of each side.

c. Find the slope of each side.

d. Find the equation of a line perpendicular to $\overline{BC}$ through point B.

6. When the members of a fife-and-drum band are arranged in a rectangular formation, the number of players in each row is 7 less than the number of rows.

a. Write an equation giving the number of players as a function of the number of rows.

b. Find the number of rows if there are 144 members in the band.

c. If there are 120 band members, how many people are in each row?

d. Find the number of players in each row and the number of rows if there are 98 members in the band.

7. A 10-foot-long ladder leans against a building and makes an angle of 68° with the ground.

a. How far from the building is the base of the ladder?

b. How far off the ground does the ladder touch the building?

c. Suppose a ladder that is twice as long is placed against the building with the base of the ladder eight feet from the building. What angle does the ladder make with the ground? With the building?

8. Consider the following data table.

x	1	2.2	3	1.7	3.7	2.6	1.8	4.3	4.5
y	2.3	2.3	2.9	2.9	3.5	3.6	2.4	3.9	5.1

a. Is this set of data reasonably modeled by a line? Explain.

b. Find the equation of the regression line for the data and graph it on the scatterplot.

c. Find the largest residual.

9. Consider $\triangle ABC$ with vertices $A(-3, 1)$, $B(-7, -2)$, and $C(1, -2)$. Find the vertices of the image of $\triangle ABC$ under each transformation.

a. Reflection across the x-axis

b. Translation with components 2 and -3

c. Size transformation $\begin{bmatrix} 2.5 & 0 \\ 0 & 2.5 \end{bmatrix}$

10. Consider the vertex-edge graph shown below.

a. Find four different paths from A to C.

b. What is the shortest path from A to C?

c. Find a minimal spanning tree.

d. What is the shortest path from A to C on your minimal spanning tree?

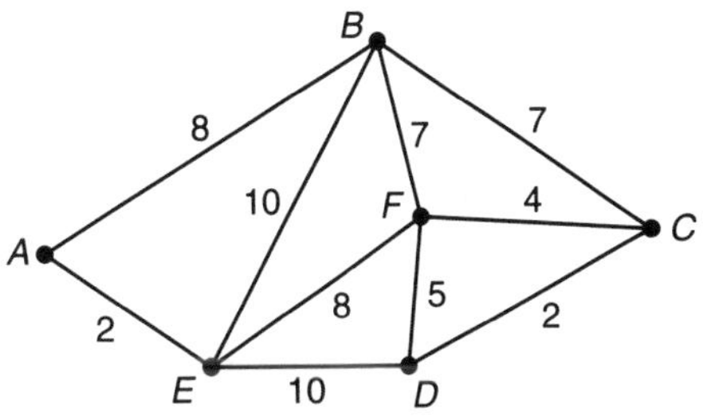

Exercise Set 5

1. Evaluate each expression if $x = 1$, $y = 3$, $a = -1$, and $b = 2$.

a. $a^2x + b^3y$ **b.** $a^2x - by$ **c.** $a^2(x + by)$

d. $\frac{(x + ay)^2}{ax - b^2}$ **e.** $\frac{x^3}{by + a^3}$ **f.** $(ax + by)(x - xby)$

2. Solve each linear system by reasoning with the symbols themselves. Check each result.

a. $2x + y = 8$
$3x - 2y = 5$

b. $3x + 5y = 3$
$x + 2y = 13$

c. $3x - 8y = 7$
$x + 2y = -7$

3. At a rock concert, tickets sold for \$15 and \$25. If the 2,000-seat arena was sold out and \$38,000 were reported in ticket sales, how many seats at each price were sold?

4. Solve each quadratic equation by reasoning with the symbols themselves. Express all answers in simplified radical form.

a. $3x^2 + 7 = 16$ **b.** $2x^2 - 9 = 13$ **c.** $15 + 4x^3 = 39$

d. $16 + 5x^2 = 74$ **e.** $x^3 - 18 = 46$ **f.** $\frac{2}{5}x^2 - 3 = -2$

5. Write each expression in an equivalent, simplified form using only positive exponents. Assume all variables are nonzero.

a. $x^{-3} \cdot y^2$ **b.** $t^{-4} \cdot t^{-1}$ **c.** $\frac{a^{-2}}{a^2}$ **d.** $(x^3)^2$

e. $(6x^3)^3$ **f.** $-2x^{-2}$ **g.** $(6x^3)(2x^2)$ **h.** $\frac{2y^6x^4}{12yx^{-1}}$

6. $\triangle ABC$ has vertices $A(4, -1)$, $B(6, 2)$, and $C(9, 0)$.

a. Find the perimeter of $\triangle ABC$.

b. Find the area of $\triangle ABC$.

c. $\triangle A'B'C'$ is the image of $\triangle ABC$ under a size transformation of magnitude 3 and center at the origin.

- What is the perimeter of $\triangle A'B'C'$?
- What is the area of $\triangle A'B'C'$?

d. $\triangle A_1B_1C_1$ is the image of $\triangle ABC$ under a reflection across the line $y = x$.

- What is the perimeter of $\triangle A_1B_1C_1$?
- What is the area of $\triangle A_1B_1C_1$?

7.Consider the data table below.

x	2.2	3.1	4.3	2.7	5.2	3.5	5.7	2.5	5.5
y	3.9	5.0	8.6	4.5	9.8	5.5	7.5	4.2	10.2

a. Is this set of data reasonably modeled by a line? Explain. If so, find the equation of the least squares regression line.

b. Find the correlation coefficient.

c. Which point appears to be the most influential? Will the correlation coefficient increase or decrease if it is removed?

8. The length of a square lot is increased by 10 feet.

a. Write an equation giving the new area of the lot.

b. Estimate the original width of the lot if the new area is 600 square feet.

c. If the new perimeter is 180 feet, what is the new length of the lot?

d. If the new area is 875 square feet, what is the new perimeter of the lot?

9. In order to make a smooth landing, the pilot of an airplane must gradually descend toward the runway. Suppose a small airplane is approaching the Cedar Rapids airport at an altitude of 3,000 feet.

a. If the pilot wants the path of the plane to make a 5° angle with the ground, at what ground distance from the airport must she start descending? How far will she fly before touching down?

b. If the pilot starts descending 10,000 feet from the airport, what angle will the plane's path make with the horizontal?

10. Consider the quadratic function defined by $y = -x^2 + 5x - 6$.

a. Find the x- and y-intercepts of the graph of the function.

b. Find the equation of the symmetry line of its graph.

c. Which way does the graph open? Explain your answer.

d. How many solutions does the equation $-12 = -x^2 + 5x - 6$ have? Explain your reasoning.

Exercise Set 6

1. Sketch a graph of each function.

a. $y = 2x^2$ **b.** $y = x^2 + 2$ **c.** $y = 2^2 + x$

d. $y = (2x)^2$ **e.** $y = -x^2 + 2$ **f.** $y = \frac{2}{x^2}$

2. Use matrix methods to solve this system of equations; then solve the system a second way by reasoning with the symbols themselves.

$$4x - 3y = 5$$
$$-2x + 9y = -1$$

3. Solve each equation using symbolic reasoning.

a. $6x^2 = 24$ **b.** $(x + 2)^2 = 81$

c. $x^2 + 9 = 4$ **d.** $2x^3 = -16$

e. $x(x + 4) = 12 + x(x + 5)$ **f.** $(3x)^3 = 64$

4. A boat rental company charges \$200 per day plus \$2.50 for every gallon of gas used.

a. What equation shows how to calculate the daily bill y as a function of gallons of gasoline used x?

b. Express the relationship using *NOW* and *NEXT*.

c. Your bill for one day of boat rental was \$230. How many gallons of gas did you use?

d. If you only have \$250 for a day's outing, how many gallons of gas can you buy?

e. If your bill is \$241.75, what does that tell you about the way they charge for gas use?

5. Write each expression in an equivalent and simplified form; then find the decimal equivalent of each number. Assume $x > 0$.

a. $\sqrt{4} \cdot \sqrt{64}$ **b.** $\sqrt{6} \cdot \sqrt{32}$ **c.** $\left(\frac{36}{49}\right)^{\frac{1}{2}}$ **d.** $2\sqrt{x^6} \cdot 4\sqrt{16}$

e. $2\sqrt{18} \cdot 3\sqrt{42}$ **f.** $\left(6\sqrt{2x}\right)^2$ **g.** $\frac{5\sqrt{15}}{\sqrt{5}}$ **h.** $24^{\frac{1}{3}}$

6. $\triangle ABC$ has vertices $A(-3, 2)$, $B(-6, -3)$, and $C(-9, -3)$. Find the vertices of the image of $\triangle ABC$ under each of the following transformations.

a. A 90° clockwise rotation centered at the origin

b. Reflection across the line $y = -x$

c. Translation with components 3 and -5

d. Size transformation with magnitude 4 and center at the origin

Exercise Set 6

7. Given the quadratic function defined by $y = x^2 - 11x + 10$,

a. Find the x- and y-intercepts of its graph.

b. Find the equation of the symmetry line of its graph.

c. Which way does the graph open? Explain your answer.

d. Sketch the graph of $y = x^2 - 11x + 10$.

8. Find the lengths of the sides and the measures of the angles in each right triangle below.

a.

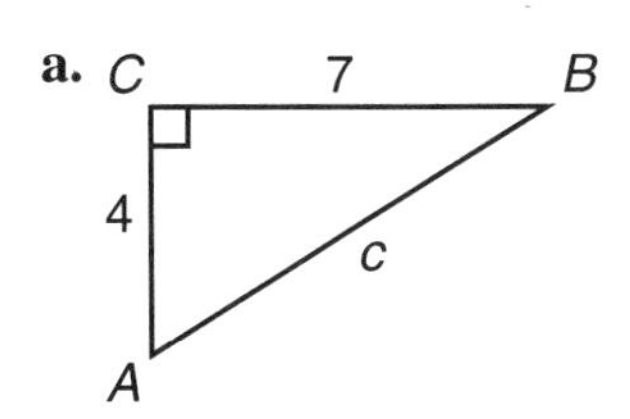

b.

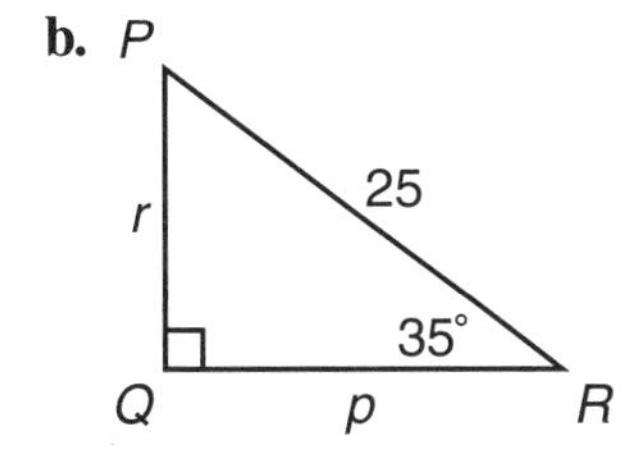

9. A Ferris wheel that is 40 feet in diameter makes one revolution every two minutes.

a. What is the angular velocity in radians per second?

b. What is the linear velocity, in feet per second, of a seat on the outer edge of the wheel?

10. A carnival game is set up so that the player has a 5% chance of winning a prize worth \$1, a 5% chance of winning a prize worth \$10, and a 90% chance of winning nothing. What is the fair price to charge to play this game?

Exercise Set 7

1. On a 130-mile trip from Columbus to Wheeling, at 10 miles and 10 minutes out of Columbus, Josiah sets his cruise control to 60 miles per hour.

a. What equation shows how to calculate distance y which Josiah has traveled as a function of the number of hours driven x?

b. What equation using *NOW* and *NEXT* shows how the distance traveled changes for each additional hour driven x?

c. How long will it take Josiah to travel 125 miles?

d. How far has he driven after 70 minutes?

e. Suppose Jenny, leaving at the same time as Josiah, takes 15 minutes to travel those first 10 miles. If she sets her cruise control to 65 miles per hour, at what time will she catch up with Josiah?

f. When she catches up with Josiah, how far is she from Columbus?

2. Solve each linear system by reasoning with the symbols themselves. Check each result.

a. $y = x + 2$
$2x + y = 11$

b. $3x - 2y = 1$
$4y = 7 + 3x$

c. $x - 5y = 2$
$2x + y = 4$

3. The intensity of sound, when measured in watts per square meter, is inversely proportional to the distance squared from the source. The distance is in meters.

a. Express the intensity I in terms of the distance in meters m from the source and a constant of proportionality w.

b. What is w if $I = 80$ watts per square meter when $m = \frac{1}{2}$ meter?

c. What is the intensity of the sound at a distance of 2.5 meters?

d. How far from the source is the intensity 3.2 watts per square meter?

4. Before last weekend's hiking trip, Juanita put together a trail mix of 3 pounds of peanuts and raisins for an energy snack. The peanuts cost \$4.25 per pound, and the raisins cost \$3.50 per pound. Together, the whole mix cost \$12.

a. Write an equation showing the relationship among peanuts, raisins, and the weight of the trail mix.

b. Write an equation showing the relationship among peanuts, raisins, and the cost of the trail mix.

c. How many pounds of peanuts did Juanita use, and how many pounds of raisins did she use?

Exercise Set 7

5. Write an equivalent and simplified form of each expression using only positive exponents. Assume all variables are nonzero.

   a. $(x^3y^{-2})^2$ b. $x^3 \cdot x^2y$ c. $\frac{x^4}{x^{-2}}$

   d. $(3x^{-1})^4 \cdot 2xy^2$ e. $(2x^4)^{-2}$ f. $3x^{-4} \cdot 6x^3$

6. $\triangle ABC$ has vertices $A(-2, 3)$, $B(-5, 4)$, and $C(x, 3)$.

   a. Find the coordinates of point C so that $AB = BC$.

   b. Find the length of each side of $\triangle ABC$.

   c. Is $\triangle ABC$ a right triangle? Justify your reasoning.

   d. Find the equation of the line containing point B and perpendicular to $\overline{AC}$.

7. Consider $A = \begin{bmatrix} -2 & 3 & 1 \\ 4 & -3 & 0 \end{bmatrix}$, $B = \begin{bmatrix} 1 & 2 & 1 \\ 3 & -5 & 2 \end{bmatrix}$, and $C = \begin{bmatrix} 2 & 1 \\ 3 & -2 \\ -3 & 1 \end{bmatrix}$.

   a. Find $A + B$. b. Find $A - B$.

   c. Find CB. d. Find $3C$.

8. Suppose that Pizza Quick advertises that their pizza costs 10 cents a square inch.

   a. What is the price for an 8"-diameter pizza?

   b. Write a general equation for the price of a pizza that has a diameter of d inches.

   c. How large a pizza did Sara get if she paid \$11.31 (exclusive of tax) for the pizza?

   d. How do the costs compare for an 8" pizza and a 16" pizza?

9. Hikers, 24 meters from the base of a sheer cliff, sight an eagle's nest as shown at the right. How far is the eagle's nest from the top of the cliff?

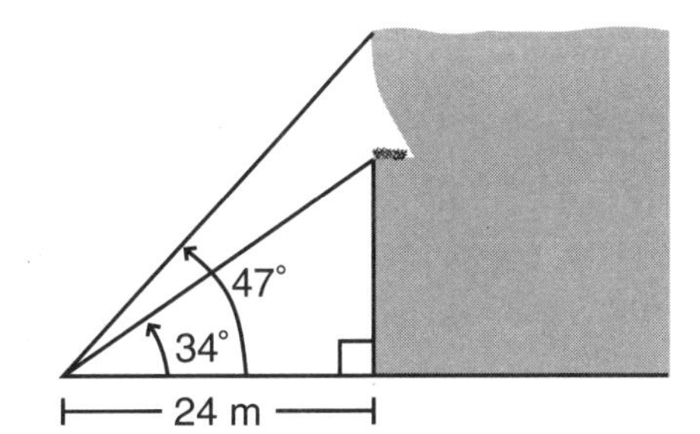

10. Each time Sandy is at bat, she has a 0.425 chance of getting a hit. What is the probability she doesn't get a hit on either of her next two at bats?

Exercise Set 8

1. A family that lives in the center of a large city plans to rent a car for a one-day trip to visit relatives. One car rental company charges \$25 per day and provides 100 free miles, then charges 40 cents per mile for any miles beyond the first 100.

   a. What equation shows how to calculate the rental fee y as a function of the number of miles driven beyond the first 100 free miles x?

   b. What equation using *NOW* and *NEXT* shows how the rental bill grows as each additional mile beyond the first 100 free miles is driven?

   c. Find the distance driven if the rental bill was \$59.40.

   d. Find the cost if you drive 145 miles.

   e. Find the range of distances possible if the rental bill is to remain below \$55.

   f. A second car rental company also offers 100 free miles, but charges \$10 per day and 50 cents per mile. For what distances is the second rental company a better choice?

2. Solve each system of equations. Check each result.

   a. $y = x^2$, $y = -x + 6$ **b.** $y = -x^2 + 5$, $x + 2y = 4$ **c.** $x^2 + 4x - y = -4$, $2x + 3y = 12$

3. Given the function $y = x^2 + 6x - 16$:

   a. Find y if $x = -3$.

   b. Find the x-intercepts of the graph of this function.

   c. Sketch the graph of this function.

4. Write each expression in an equivalent and simplified form. Assume $x > 0$ and $y > 0$.

   a. $\sqrt{49}\sqrt{169}$ **b.** $\sqrt{72}\sqrt{24}$ **c.** $\left(\frac{64}{144}\right)^{\frac{1}{2}}$ **d.** $3\sqrt{42} \cdot 2\sqrt{51}$

   e. $\sqrt{\frac{27}{3}}$ **f.** $x^3\sqrt{4x^6y}$ **g.** $\sqrt{\frac{6x^3y}{24xy^3}}$ **h.** $\sqrt{(x + 7)^2}$

5. Quadrilateral *WXYZ* is a rectangle represented by the matrix

$$WXYZ = \begin{bmatrix} 2 & 6 & 8 & x \\ 4 & 0 & 2 & y \end{bmatrix}.$$

   a. Find the coordinates of point Z.

   b. Verify that *WXYZ* is a rectangle.

   c. Find the area of rectangle *WXYZ*.

Exercise Set 8

6. The base b and height h of triangles with an area of 24 cm^2 are related by the formula $h = \frac{48}{b}$.

a. If the base of the triangle with area 24 cm^2 is 12 cm, what is the height?

b. If the height of one of these triangles is 10 cm, what length is the base?

c. Sketch a graph of this relationship.

7. A 26-inch bicycle wheel is turning at a rate of 3 revolutions per second.

a. How far will the bike travel in one second?

b. Write an equation that expresses distance traveled in terms of time in seconds.

c. How long will it take to travel 100 yards?

d. What is the angular velocity of the rotating wheel in radians per second?

8. In a data-collecting project at Arbor High School, the width and length of the right feet of 100 juniors were measured in inches.

a. Would you expect the correlation to be positive or negative? Explain your reasoning.

b. If the slope of the regression line for the collected data is 2.74, what would the slope mean in this context?

9. Write a matrix representation for each transformation, then find the image of $P(2, 3)$ in each case.

a. Reflection across the y-axis

b. Size transformation with scale factor 0.5 and center at the origin

c. Reflection across the line $y = -x$

d Counterclockwise rotation of 270° about the origin

10. Consider the vertex-edge graph below.

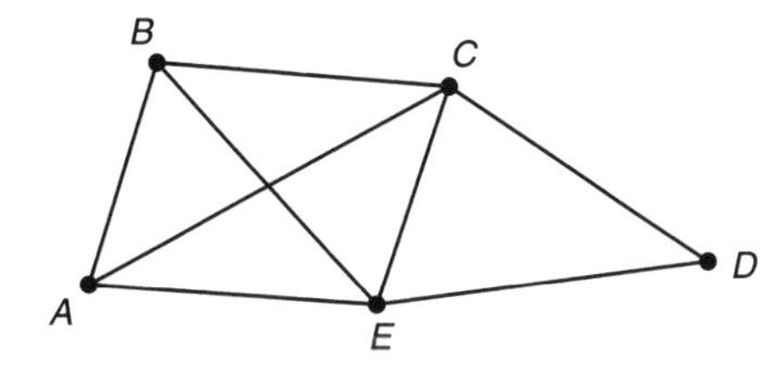

a. Create an adjacency matrix for the graph.

b. How many paths of length 3 are there from A to D?

Exercise Set 9

1. The National Honor Society at Dawson High School is selling school notebooks to raise money for a service project. The notebooks cost \$1.80 each and there is a \$48 design setup charge. They plan to sell the notebooks for \$3.00 a piece.

a. Write an equation expressing profit p as a function of the number of notebooks sold n.

b. How many notebooks do they need to sell to break even?

c. If they made a profit of \$236.40, how many notebooks did they sell?

2. Rewrite each equation in the form $y = \ldots$.

a. $x = 32 + \frac{9}{5}y$ **b.** $5 = \frac{x + y}{10}$ **c.** $2(x - y) - 3(x + y) + 4 = 0$

3. Solve each linear system by reasoning with the symbols themselves. Check each result.

a. $4x - 3y = 2$
$5x - y = -3$

b. $3x - 2y = 8$
$5x + 7y = 3$

c. $5x - 2y = 1$
$15x - 6y = 7$

4. Sketch the graphs of each of the following equations.

a. $y = -2x^2 + 2$ **b.** $y = -3x^2 + 2$ **c.** $y = 3x^2 + 2$
d. $y = x^3 + 2$ **e.** $y = -x^3 + 2$ **f.** $y = x^3 - 2$

5. Solve each quadratic equation using symbolic reasoning.

a. $8x^2 - 7 = 41$ **b.** $9x^2 - 1 = 80$ **c.** $2x^2 + 3 = 21$

d. $4x^2 - 27 = 3$ **e.** $5 - 3x^2 = -32$ **f.** $\frac{1}{3}x^2 + \frac{2}{3} = 1$

6. Find values for the variables in each matrix equation.

a. $\begin{bmatrix} x & y \\ 2 & 3 \end{bmatrix} + \begin{bmatrix} 3 & -4 \\ 6 & 9 \end{bmatrix} = \begin{bmatrix} 10 & -10 \\ z & 12 \end{bmatrix}$

b. $\begin{bmatrix} 6 & 2 \\ 9 & x \end{bmatrix} \begin{bmatrix} y \\ 3 \end{bmatrix} = \begin{bmatrix} 42 \\ 60 \end{bmatrix}$

c. $\begin{bmatrix} 2 & -5 \\ 1 & -4 \end{bmatrix} + 3\begin{bmatrix} 2x & 1 \\ y & -7 \end{bmatrix} = \begin{bmatrix} 14 & -2 \\ -11 & z \end{bmatrix}$

Exercise Set 9

7. Jason throws a rock straight up in the air. The height (in meters) of the rock after t seconds is given by $h = -4.9t^2 + 20t + 1.2$.

   a. What do the coefficients -4.9 and 20 represent in this context? What does the term 1.2 represent?

   b. How high is the rock after 1.3 seconds?

   c. When is the rock 10 meters above the ground?

   d. When does the rock reach its highest point and how high is it?

   e. How long is the rock in the air?

8. Find the area of each figure below.

   a.

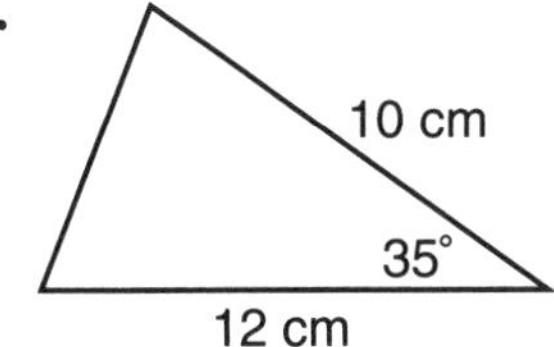

   b.

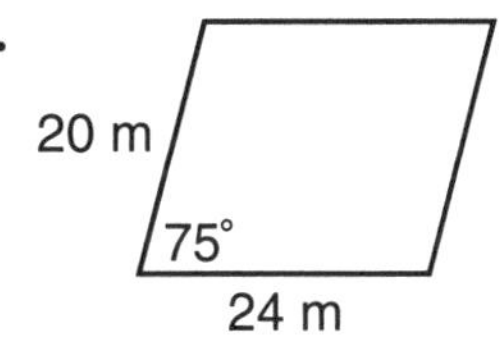

   c.

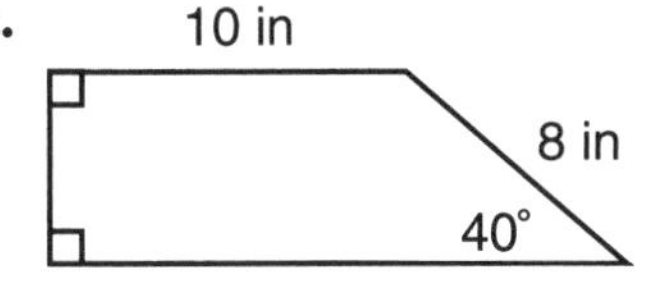

   d.

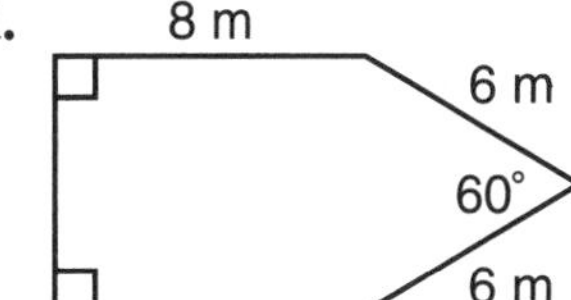

9. Evaluate each expression if $x = -4$, $y = 3$, and $z = \frac{1}{2}$.

   a. x^2y^3z b. $4(2y - x)^2$ c. $|x - y| + |4z - x|$

   d. $2(2x + 4) - 6(z - 1)$ e. $\sqrt{x^2y^6}$ f. z^{-1}

10. The backs of the labels of 40% of the bottles of a certain brand of water have a coupon for a movie discount. Suppose these labels are placed randomly on the bottles.

    a. Make a probability distribution for the number of bottles you would have to buy before getting a coupon.

    b. What is the probability that the fourth bottle you purchase will be the first to have the coupon?

    c. Would a rare event have occurred if six bottles were purchased to get the coupon? Explain.

Exercise Set 10

1. Consider the matrix representations below of a transformation T and quadrilateral $ABCD$.

$$T = \begin{bmatrix} 0 & 1 \\ -1 & 0 \end{bmatrix} \qquad ABCD = \begin{bmatrix} 0 & -2 & 1 & 3 \\ 0 & 6 & 7 & 1 \end{bmatrix}$$

a. Find the perimeter and area of quadrilateral $ABCD$.

b. Find the image $A'B'C'D'$ of quadrilateral $ABCD$ under the transformation T.

c. What is the perimeter and area of quadrilateral $A'B'C'D'$?

d. What type of transformation does T represent?

e. Write a matrix representation for a size transformation with magnitude 1.5 and center at the origin.

f. Find the image $A_1B_1C_1D_1$ of quadrilateral $ABCD$ under the size transformation in Part e.

g. What are the perimeter and area of quadrilateral $A_1B_1C_1D_1$?

2. Solve each equation.

a. $\sqrt{x+4} = 7$ **b.** $x^3 + 10 = 9$ **c.** $(x + 10)^3 = 8$

d. $\sqrt{2x} + 6 = 10$ **e.** $x^{\frac{1}{3}} - 5 = 1$ **f.** $\left(x^{\frac{1}{2}}\right)^2 = 2x + 8$

3. Use matrix methods to solve this system of equations; then solve the system a second way by reasoning with the symbols themselves.

$$x - 5y = 7$$
$$4x + 9y = 28$$

4. Write each expression in an equivalent and simplified form with positive exponents. Assume all variables are nonzero.

a. $y^3 \cdot y^{-2}$ **b.** $q^{-2} \cdot q^2$ **c.** $\dfrac{t^{-3}}{t^{-2}}$ **d.** $(x^2)^3$

e. $(a^{-2})^3$ **f.** $2x^{-2} \cdot 3x$ **g.** $\dfrac{6x^2}{x^4y^{-2}}$ **h.** $b^3(b^4)^2 + b^{11}$

5. A 12-foot long ladder leans against a building and makes an angle of 70° with the ground.

a. How far from the building is the base of the ladder?

b. How high up the building does the ladder reach?

c. Repeat Part a if the ladder is 8 feet long.

d. Calculate, in three different ways, how high up the building the 8-foot ladder reaches.

Exercise Set 10

6. Quadrilateral $ABCD$ has vertices $A(-3, -3)$, $B(-2, 0)$, $C(4, 4)$, and $D(0, -2)$.

   a. What kind of quadrilateral is this? Explain.

   b. Find the coordinates of the intersection point for the two diagonals.

   c. At what angle(s) do the diagonals meet?

7. Consider the data in the table below.

x	1.2	2.3	2.7	1.7	2.4	4.7	1.6	4.2	3.3
y	8.2	6.5	5.9	5.4	4.4	2.9	5.6	3.7	3.0

   a. Is this set of data reasonably modeled by a line? Explain.

   b. Find the equation of the regression line for the data plot.

   c. Find the correlation coefficient.

   d. Find the residuals for $x = 1.2, 4.7$, and 2.3.

8. Model each situation with an equation. Then sketch the graph of the model and answer the question.

   a. The number of pepper plants in a row in Nicole's garden varies inversely as the space between them. If the plants are spaced 12 inches apart, 50 plants fit in a row. How many can fit if the plants are spaced 25 inches apart?

   b. The time required for Nathan to drive between two cities varies inversely as the speed of his truck. It takes him three hours driving at 55 miles per hour. How long does it take if he drives at 60 miles per hour?

9. A bag contains 2 red marbles, 3 green marbles, and 4 white marbles. Two marbles are randomly selected with replacement. Find the probability of each event.

   a. The first is green and the second is white.

   b. Both are white.

   c. Both are green.

   d. One is green and one is white.

10. Given $\triangle ABC$ with vertices $A(-2, 1)$, $B(-3, 4)$, and $C(1, 3)$, find the vertices of the image of $\triangle ABC$ under each transformation.

   a. Reflection across the x-axis

   b. 180° rotation about the origin

Exercise Set 11

1. Suppose you discover that a relation among variables r, k, s, t, and u has the form $r = \frac{kst}{u}$.

a. Find r if $k = 3$, $s = 5$, $t = 8$, and $u = 0.5$.

b. Find t if $r = 300$, $k = 5$, $s = 20$, and $u = 15$.

c. Find u if $r = 2.4$, $k = 5$, $s = 30$, and $t = 16$.

d. Express s in terms of r, k, t, and u.

e. Solve for u in terms of r, k, s, and t.

2. Solve each linear system by reasoning with the symbols themselves. Check each result.

a. $33x - 9y = 300$
$11x - 3y = 100$

b. $2x - y = 135$
$7x + 4y = 15$

c. $x + y = 16$
$y = x + 2$

3. Draw a scatterplot of 10 data points that illustrates each condition.

a. Strong negative association

b. Weak positive association

c. Modeled well by a linear equation

d. Modeled well by an exponential equation with base $b > 1$

e. Modeled well by a quadratic power rule

4. Solve each quadratic equation using symbolic reasoning.

a. $2x^2 - 5 = 15$

b. $7 = 6x^2 - 19$

c. $25x^2 - 31 = 8$

d. $4 = 13 - 16x^2$

e. $-5 = 13 - 27x^2$

f. $\frac{-1}{2}x^2 + 3 = -2$

5. Suppose $A = \begin{bmatrix} -2 & 1 \\ 4 & 3 \end{bmatrix}$, $B = \begin{bmatrix} 5 & -2 \\ -3 & 1 \end{bmatrix}$, and $C = \begin{bmatrix} 1 & 5 & 2 \\ 3 & 9 & -1 \end{bmatrix}$.

a. Find AB and BA. What do you notice?

b. Find AC and CA. What do you notice?

c. Find $A + B$ and $B - A$.

d. Find $D = \begin{bmatrix} x \\ y \end{bmatrix}$ such that $AD = \begin{bmatrix} 8 \\ -6 \end{bmatrix}$.

e. Find the inverse of B.

Exercise Set 11

6. Consider the function $f(x) = x^2 + 2x - 8$.

a. Describe the domain and range of $f(x)$.

b. Find $f(-3)$.

c. For what value of x is $f(x) = 40$?

d. Find the x-intercepts of the graph $f(x)$.

7. Suppose that the time needed to complete a job varies inversely with the number of people working on the job. It takes 8 hours for 3 people to complete the job.

a. Write a modeling equation for this situation.

b. Sketch a graph of the equation from Part a.

c. How long will it take to complete the job if only 6 people are working?

d. How many people will be needed if the job must be completed in 90 minutes?

8. The matrix below gives weights (distances) for a vertex-edge graph G.

$$G = \begin{array}{c} \\ A \\ B \\ C \\ D \\ E \end{array} \begin{array}{c} \begin{array}{ccccc} A & B & C & D & E \end{array} \\ \begin{bmatrix} - & 2 & 2 & - & 3 \\ 2 & - & - & 5 & 3 \\ 2 & - & - & 7 & 3 \\ - & 5 & 7 & - & 4 \\ 3 & 3 & 3 & 4 & - \end{bmatrix} \end{array}$$

a. Draw graph G.

b. Find a minimal spanning tree for G. What is the length of the spanning tree?

c. Solve the Traveling Salesperson Problem for graph G.

9. Two points in a data set are (3, 5) and (6, 11). The least squares regression line residuals are 0.65 for (3, 5) and 2.8 for (6, 11).

a. Find the equation of the regression line.

b. Another point in the data set is (7, 10). What is its residual?

10. The vertical distance between the second and third floors of a department store is 28 feet.

a. How long must an escalator be if the angle it makes with the floor is 40°?

b. What is the horizontal distance covered by the escalator?

Exercise Set 12

1. Evaluate each expression if $x = -1$, $y = -2$, $z = 3$, $a = 4$, and $c = 2$.

a. $a^2x + y$

b. $a^2cx - z^3y$

c. $a^2(c^2x + z^3y)$

d. $\dfrac{(x + ayz)^2}{ax + c^2z}$

e. $\dfrac{x^3}{cy + z^2}$

f. $(ax + zy)(cx - xy)$

2. The volume of a right circular cone is given by the equation $V = \frac{1}{3}\pi r^2 h$, where r is the radius of the base and h the altitude of the cone.

a. Express h as a function of V and r.

b. Express r as a function of V and h.

c. If h is tripled and r remains unchanged, how is V affected?

d. If r is quadrupled and h remains unchanged, how is V affected?

3. Solve each system of linear inequalities.

a. $x + y \geq 4$
$x - 2y \leq 6$

b. $y \leq -2x + 1$
$-3x + y \leq 6$

c. $x \geq 4$
$x - y \geq 0$

4. Rectangle $A'B'C'D'$ is the image of rectangle $ABCD$ under a size transformation of magnitude 3 with center at the origin.

a. If the length of $\overline{AB}$ is 6, what is the length of $\overline{A'B'}$?

b. If $A'B' = 21$, find AB.

c. If the area of quadrilateral $A'B'C'D'$ is 72 square units, what is the area of quadrilateral $ABCD$?

d. The slope of $\overline{C'D'} = 23$. Find the slopes of $\overline{CD}$ and $\overline{BC}$.

5. Write each expression in an equivalent and simplified form using positive exponents. Assume all variables are nonzero.

a. $x^{-2} \cdot x^3y^{-1}$

b. $(4x^3)^4$

c. $\dfrac{-2x^3}{4x^{-1}}$

d. $\left(\dfrac{3x^4}{6}\right)^{-2}$

e. $(-2y^4)^2(x^3y^2)$

f. $\dfrac{4x^2y^4z^9}{(2xy^3)^2}$

6. Find exact solutions to each quadratic equation.

a. $x^2 + 4x - 5 = 0$

b. $x^2 - 2x - 3 = 0$

c. $x^2 - 5x - 6 = 0$

d. $-x^2 - 2x + 15 = 0$

Exercise Set 12

7. $\triangle ABC$ has vertices $A(2, 6)$, $B(0, 1)$, and $C(-3, 4)$.

a. Find the slope of side $\overline{BC}$.

b. What kind of triangle is $\triangle ABC$?

c. Find the equation of the altitude from point A to $\overline{BC}$.

d. In what point does the altitude from point A to $\overline{BC}$ intersect $\overline{BC}$? How is the point related to points B and C?

8. Find the missing measures in each figure below.

a.

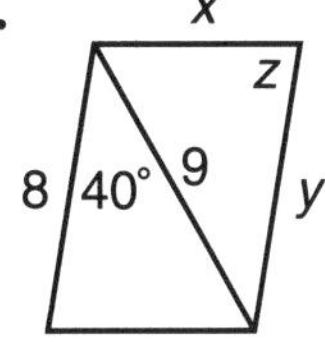

parallelogram

b.

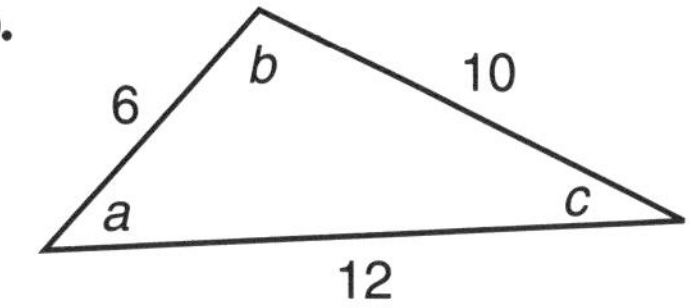

9. Consider the quadratic function defined by $y = x^2 + 4x - 5$.

a. Find the x- and y-intercepts of its graph.

b. Find the equation of the symmetry line of its graph.

c. Which way does the graph open? Explain your answer.

d. Sketch the graph of $y = x^2 + 4x - 5$.

10. The box plot below shows the likely sample outcomes for random samples of size 50 taken from a population in which 20% have a certain characteristic.

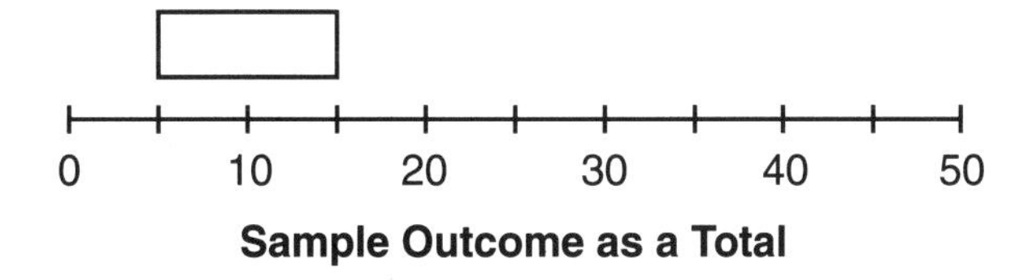

At Community High School, 20% of the students are seniors. However, of the 50 students in the Honor Society, 26% are seniors. Can you conclude that seniors are more likely to be members of the Honor Society? Explain your reasoning.

Exercise Set 13

1. Sketch the graph of each equation. For which graphs were you able to make a quick sketch by locating x- and/or y-intercepts?

a. $3x - 4y = 12$ **b.** $3xy = 12$ **c.** $y = 4 \cdot 3^x$

d. $y = -x^2 + 4$ **e.** $y = \frac{4}{x^2}$

2. Solve each equation for x by reasoning with the symbols themselves.

a. $3(x + 4) - 2(x - 5) = 10$ **b.** $(x + 2)^2 = 16$

c. $x^2 = 7x - 10$ **d.** $4x^{\frac{1}{2}} + 6 = 15$

e. $4x(x - 2) = 8 - 2 + (4x - 3)x$

3. The current I in amps drawn by a circuit is the sum of the currents drawn by each appliance of the circuit. Each current is given by the equation $I_n = \frac{V}{R_n}$ where V is the voltage of the circuit and R_n is the resistance in ohms.

a. Write an equation representing the current I drawn when three appliances with resistances R_1, R_2, and R_3 are in use simultaneously on a 120-volt circuit.

b. Suppose a microwave with a resistance of 30 ohms and a toaster oven with a resistance of 20 ohms are operating at the same time on a circuit, what current is used if the voltage is the standard 120?

c. Most circuits have 15-amp fuses installed so that the circuit opens if the current exceeds 15 amps. If a third appliance with a resistance of 20 ohms is added to the circuit in Part b, what happens?

4. A national poll announces that it found that 63% of voters are in favor of some issue. The poll has a margin of error of 3%. Does this mean that the percentage of all voters in favor of the issue is between 60% and 66%? Explain your reasoning.

5. Write each expression in an equivalent and simplified form. Assume all variables are positive.

a. $\sqrt{169x^2}\sqrt{64x^6}$ **b.** $3\sqrt{128t^5}$ **c.** $\left(\frac{36x^3}{64x}\right)^{\frac{1}{2}}$ **d.** $\sqrt[3]{48}$

e. $4\sqrt{72} \cdot 2\sqrt{86}$ **f.** $\sqrt[3]{81x^{12}y^8}$ **g.** $\left(\frac{8}{27}\right)^{\frac{1}{3}}$ **h.** $\left(4x^3\right)^{\frac{3}{2}}$

6. $\triangle ABC$ has vertices $A(-5, 1)$, $B(-9, 1)$, and $C(-1, -5)$. Find the intersection of the line containing point A and the midpoint of $\overline{BC}$ and the line containing point B and the midpoint of $\overline{AC}$.

Exercise Set 13

7. Find exact solutions to each quadratic equation.

a. $x^2 - 11x + 10 = 0$

b. $x^2 + 16x + 28 = 0$

c. $x^2 - 7x - 18 = 0$

d. $-x^2 + 10x + 24 = 0$

8. Surveyors Alice and June are standing 900 feet apart on a straight, flat road in a new housing development. A helicopter hovers in the air halfway between them.

a. If June measures the angle of elevation of the helicopter as 30°, what is the height of the helicopter?

b. If the helicopter, remaining at the same height in Part a, flies toward Alice, at what distance from her would Alice measure an angle of elevation of 50°?

c. At what distance from Alice would June measure an angle of elevation of 25°?

9. Given the quadratic function defined by $f(x) = -x^2 - 2x + 15$:

a. Which way does its graph open? Explain your answer.

b. Find the equation of the symmetry line of the graph.

c. Find all x for which $f(x) = 0$.

d. Find the maximum or minimum value of the function.

e. Sketch the graph of $y = f(x)$.

10. Write a matrix representation for each transformation.

a. Reflection across the y-axis

b. Size transformation with magnitude 5 and center at the origin

c. Counterclockwise rotation of 90° with center at the origin

d. Reflection across the line $y = -x$

Exercise Set 14

1. Carlos' height above a trampoline in one bounce can be modeled by the function $f(t) = -16t^2 + 24t$, where t is in seconds and $f(t)$ is in feet.

   a. Explain the meaning of the coefficients -16 and 24 of the function rule in terms of the context.

   b. Explain the meaning of $f(0.7)$. What is its value?

   c. According to this model, when did Carlos reach his maximum height above the trampoline and what was the height?

   d. How much time did Carlos spend in the air during the bounce?

2. Graph the solution set for each system of linear inequalities.

   a. $x \geq 5$
   $y \leq 3$

   b. $x + y < 5$
   $y < 2 + 4x$

   c. $y > -2x$
   $2x - 3y < 6$

3. The vertex-edge graph below models the sidewalks connecting seven buildings of a college campus. The numbers give the yardage between the main entrances to each building.

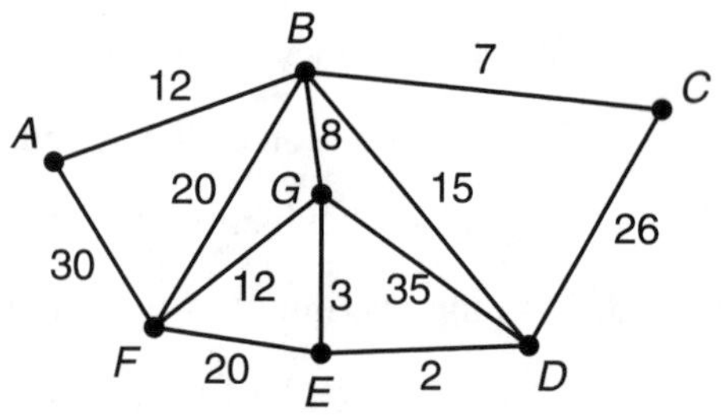

   a. After a heavy snowfall, what is the least number of sidewalk yards that need to be cleared to make each building accessible from all others? Show your sidewalk choices on a vertex-edge graph.

   b. What sidewalks need to be cleared so the campus mail delivery person could begin and end at A and reach each building without visiting any building more than once? How far will the mail delivery person walk?

4. Solve this linear system of equations.

   $$2.47x - 1.15y = -13$$
   $$-x - 13.32y = 0.27$$

5. Sketch the graphs of the three equations in each set. Describe their similarities and differences.

   a. $y = x^2 - 9$
   $y = -12 + x^2$
   $y = 0.4 + x^2$

   b. $y = 3x^2 - 10$
   $y = -5x^2 - 10$
   $y = \frac{1}{4}x^2 - 10$

Exercise Set 14

6. Write each expression in an equivalent and simplified form that does not involve parentheses.

a. $3x^2 + 2x(x - 7)$ **b.** $x - x(1 - x)$

c. $(x + 1)(x - 2) - 3(x + 4)$ **d.** $(2x + 1)^2$

e. $x^2(x + 2) + 2(x - 7)$ **f.** $-4 - 3(2x + 3) - 4(2x + 3)$

7. Quadrilateral $ABCD$ has vertices $A(-4, -4)$, $B(-5, -2)$, $C(2, 1)$, and $D(3, -1)$.

a. What type of quadrilateral is it?

b. Are the lengths of the diagonals equal? Explain.

c. At what point do the diagonals intersect?

8. Find exact solutions to each quadratic equation.

a. $x^2 + 2x - 30 = 0$ **b.** $x^2 - 4x + 2 = 0$

c. $x^2 + 5x - 8 = 0$ **d.** $-x^2 + 3x + 4 = 0$

9. Find the values of x and y in each triangle.

a.

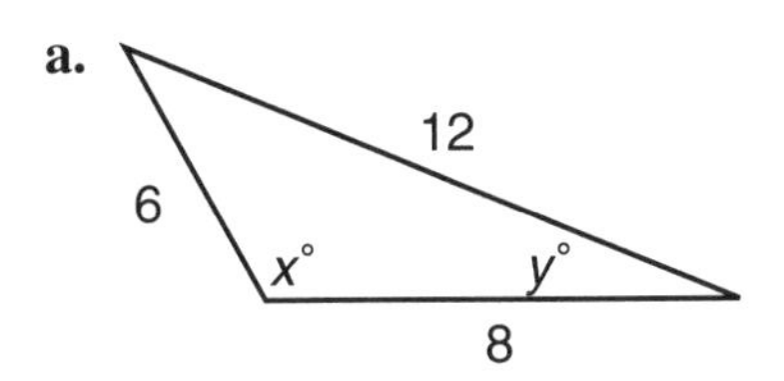

b.

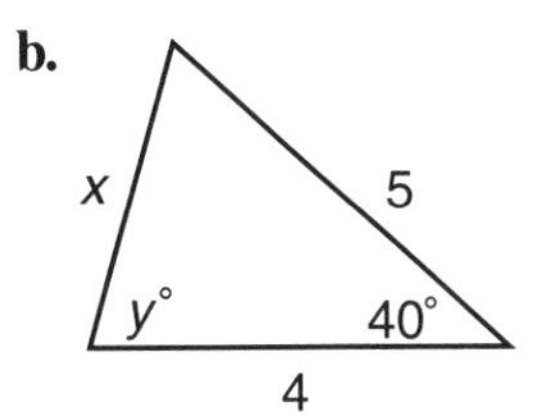

10. Consider the data in the table below.

x	3.2	1.7	1.2	3.7	4.7	5.1	3.3	5.7	1.5
y	2.0	3.0	4.8	4.2	6.0	6.5	2.3	8.0	4.0

a. Is this set of data reasonably modeled by a line? Explain.

b. Suppose (3.2, 5) and (3.3, 5.3) were misrecorded as (3.2, 2) and (3.3, 2.3). Is the correct data set reasonably modeled by a line? If so, find a modeling equation.

c. Find the residuals for $x = 5.7$ and 3.7.

Exercise Set 15

1. The figure below is a parallelogram with diagonals $\overline{BD}$ and $\overline{AC}$.

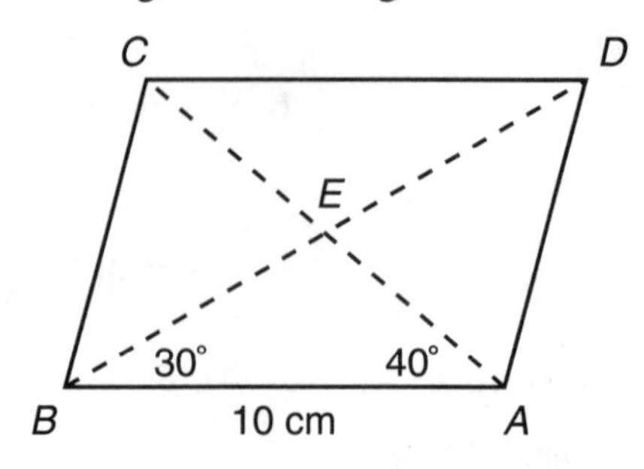

a. Find BE and AE.

b. Using the given information, results from Part a, and the fact that the diagonals of a parallelogram bisect each other, find DA.

2. Solve each equation for x in terms of the other variables. Assume all variables are positive and $k > x$.

a. $3x + b = ax - 7$

b. $\frac{c}{2x} = a + 2$

c. $xy + k = 4y + 3k$

d. $\frac{x + 1}{x + b} = a$

e. $\frac{x}{a} = \frac{b}{2x}$

f. $\sqrt{k - x} = c$

3. Sketch the graph of each inequality.

a. $y \geq x$

b. $y \leq -2$

c. $y \geq 2x + 4$

d. $3x + 2y \leq 6$

4. On one Saturday, a water park sold 400 admission tickets for a total of \$3,320. Tickets cost \$12.50 for adults and \$5.50 for children. How many tickets of each type were sold?

5. Write an equivalent, simplified form of each expression using only positive exponents.

a. $(x^2 + 4x^2)^2$

b. $(6x^{-1})^3$

c. $(x^2y^{-2})^{-3}$

d. $3x^{-2}y$

e. $(4x^2)(8x^3y)(-2x)^2$

f. $x^{-2} + x^{-2}$

Exercise Set 15

6. $\triangle ABC$ has vertices $A(-3, -1)$, $B(-5, -3)$, and $C(-7, 2)$.

a. Find the length of $\overline{CA}$.

b. Find the length of the segment determined by the midpoints of $\overline{AB}$ and $\overline{BC}$.

c. Compare the slope of $\overline{CA}$ with the slope of the segment joining the midpoints of $\overline{AB}$ and $\overline{BC}$. What can you say about the segments?

7. Find exact solutions to each quadratic equation.

a. $x^2 - 4x = 0$

b. $x^2 + 3x = 0$

c. $x^2 - 7x = 0$

d. $-x^2 + x = 0$

8. Sketch graphs of each of the following trigonometric functions for $-2\pi \leq x \leq 2\pi$.

a. $y = \sin x$

b. $y = 4\sin x$

c. $y = \sin(x) + 4$

9. A stoplight is green at the end of Tom's office driveway for 20 seconds out of every 2 minutes.

a. What is the probability that it will be green when Tom gets to it?

b. What is the probability that the next three times Tom gets to the intersection the light will be green?

c. What is the probability that he must stop when leaving work for five days in a row?

10. Write each of the following algebraic expressions in standard polynomial form.

a. $(x - 3)(x + 2)$

b. $(x - 4)^2$

c. $(x + 5)(x - 5)$

d. $(3x - 2)(x + 5)$

e. $(2x - 3)(4x - 1)$

f. $(3 - x)(x + 2)$

Exercise Set 16

1. Write each polynomial as a product of linear factors.

a. $x^2 + 8x + 12$ **b.** $x^2 + 10x + 25$ **c.** $x^2 - x - 20$

d. $x^2 - 36$ **e.** $3x^2 + 11x - 4$ **f.** $6x^2 + 5x - 4$

2. State whether you think each of the following pairs of variables will have a positive correlation, a negative correlation, or no correlation.

a. The age of a car and the value of a car

b. The size of a dress and the amount of material needed to make the dress

c. The number of pages in a novel and the length of time it takes to read the novel

d. The number of children in a family and the amount of parental attention each child receives

3. Solve each linear system. Check each result.

a. $4x + 3y = 17$
$2x + 3y = 13$

b. $y = 9 - x$
$y = 5x - 3$

c. $8x + 5y = -13$
$3x - 2y = -1$

4. Solve each equation.

a. $3x^2 = (3x + 1)(x - 5)$ **b.** $x - 2(x + 5) = 3x + 7$

c. $3(2^x) = 24$ **d.** $\frac{8}{x^2} = \frac{1}{2}$

e. $\frac{x + 2}{5} = \frac{2}{x - 1}$ **f.** $-6(2 - x) - 3(x + 4) = 10$

5. A vendor buys crates of apples and pears to sell at the Farmer's Market. The apples cost $18 per crate, and the pears cost $16.50 per crate. Each crate weighs 25 pounds. The vendor spends $3,540 to purchase 5,000 pounds of fruit. How many crates of apples and how many crates of pears did the vendor purchase?

6. Is the following statement true or false? If false, give a counterexample; if true, prove it.

Statement: If the vertex angle of an isosceles triangle has a measure of 60°, then the triangle is equilateral.

7. Consider $f(x) = x^2 + 2x - 48$.

a. State the domain and range of $f(x)$.

b. Evaluate $f(-3)$ and $f(a + 1)$.

c. Solve $f(x) = 51$.

d. Find the zeroes of $f(x)$.

e. Does $f(x)$ have a maximum or minimum value? Find it.

8. In the figure below, $l \parallel m$. Find the angle measures of all numbered angles.

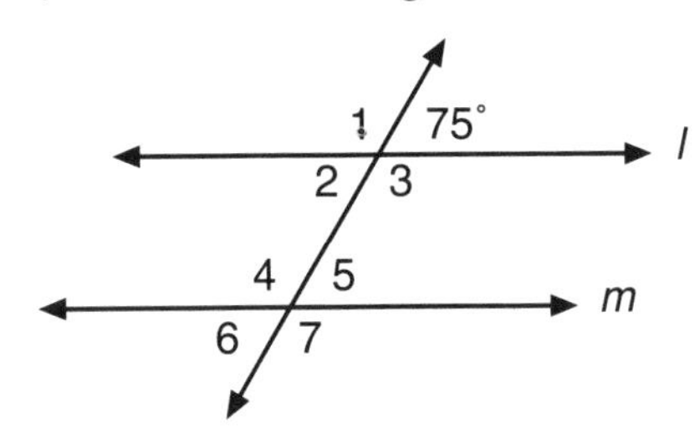

9. Find exact solutions to each quadratic equation.

a. $x^2 - x - 3 = 0$

b. $x^2 - 4x - 3 = 0$

c. $x^2 - 5x - 50 = 0$

d. $-x^2 + x + 42 = 0$

10. $\triangle ABC$ has vertices $A(1, 3)$, $B(-1, 5)$, and $C(-2, -4)$. Find the coordinates of the image of $\triangle ABC$ under each transformation.

a. Reflection across the line $y = x$, followed by a size transformation of magnitude 2 with center at the origin

b. Reflection across the x-axis, followed by a reflection across the y-axis

c. Translation with components 5 and -2 followed by a translation with components 3 and 6

Exercise Set 17

1. Solve each inequality using symbolic reasoning.

   a. $6x^2 - x - 1 > 0$ b. $6x^2 - x - 1 < 0$ c. $3x^2 - x - 10 \geq 0$

   d. $3x^2 - x - 10 \leq 0$ e. $-x^2 + 7x - 20 \leq -8$ f. $-x^2 - 7x \geq 10$

2. Sketch a graph of each function.

   a. $y = 2^x - 3$ b. $y = x^3$

   c. $y = 2\cos x$ d. $y = 12$

   e. $y = -\frac{3}{2}x + 3$ f. $2xy = 6$

3. Given that $f(x) = 3x^2 - 2x + 5$ and $g(x) = 4x - 2$, write polynomial rules for each of the following:

   a. $f(x) + g(x)$

   b. $g(x) - f(x)$

   c. $g(x) \cdot f(x)$

   d. $2g(x) - 3f(x)$.

4. Rewrite each expression in an equivalent, simplified form using only positive exponents. Assume all variables are nonzero.

   a. $x^{-2} \cdot x^4$ b. $(y^{-2})^{-3}$ c. $\frac{r^3}{r^{-2}}$

   d. $(4x^4)^2$ e. $2x^{-1} \cdot 6x^3 \cdot 2x^5$ f. $\left(\frac{x^3y}{2x^{-1}}\right)^2$

5. The vertex-edge graph at the right shows four cities in the southeast and their service by a regional airline.

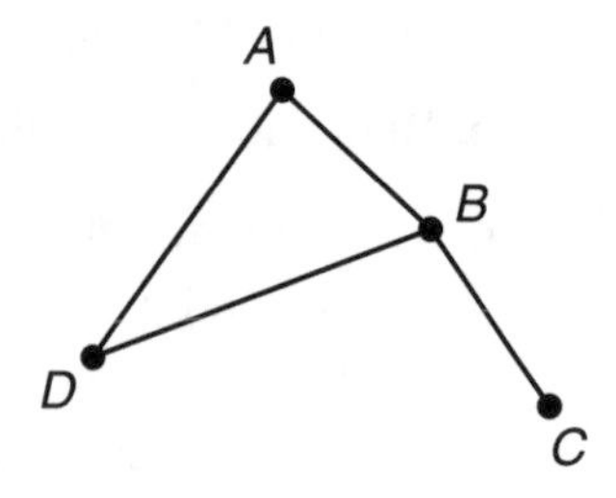

   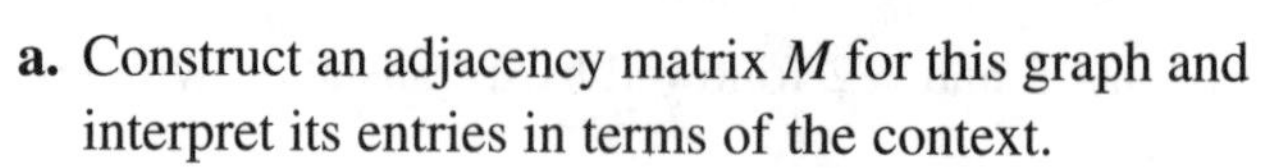

   a. Construct an adjacency matrix M for this graph and interpret its entries in terms of the context.

   b. Calculate M^2 and interpret its entries in terms of the context.

   c. Calculate M^3 and interpret its entries in terms of the context.

   d. A matrix raised to the zero power has meaning similar to that of a nonzero number raised to a zero power. If M^0 equals the indentity matrix I calculate $M^0 + M^1 + M^2$. What is the meaning of the absence of zeroes in the sum matrix?

Exercise Set 17

6. In the figure at the right, $AE = \frac{2}{3}AB$ and $AD = \frac{2}{3}AC$. Either prove each statement or provide a counterexample.

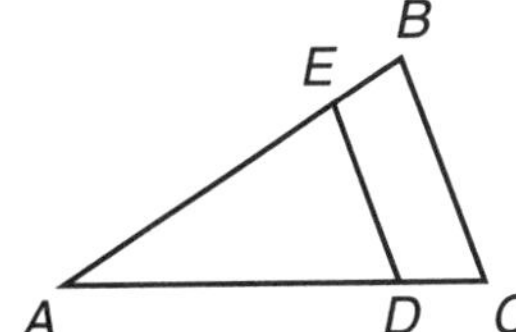

a. $\triangle ABC \sim \triangle AED$

b. $ED = \dfrac{2}{3}BC$

c. $\overleftrightarrow{ED} \parallel \overleftrightarrow{BC}$

7. Find exact solutions to each quadratic equation.

a. $x^2 + 4x - 2 = 0$

b. $x^2 + 4x + 12 = 0$

c. $(x + 2)(x - 3) = 6$

d. $x^2 = 8x$

8. A spinner is divided into three equal sections numbered 1, 2, and 3. A second spinner is divided into four equal sections numbered 2, 5, 6, and 7. You spin each spinner. You receive \$5 when both spinners fall on an odd number and their sum is greater than 7, \$3 when both spinners fall on an even number and their sum is greater than 7, and you pay \$1 in all other cases. Find the expected value of the game.

9. Tameka is standing at point A on the north side of Lake Mahopac. The west end of Lake Mahopac is about 400 meters distant, and the east end is about 800 meters distant. The angle formed by these lines of sight from point A is 125°. Find the approximate length of Lake Mahopac.

10. Consider the data in the table below.

x	3.5	4.7	2.3	3.2	1.4	3.7	2.9	4.6	1.8
y	14.5	13.7	11.2	13.6	8.8	15.0	8.8	13.4	9.9

a. Is this set of data reasonably modeled by a line? Explain.

b. Find the equation of the regression line for the data plot.

c. Find the correlation coefficient.

d. Find the residual for $x = 1.8$.

Exercise Set 18

1. Write each polynomial as the product of linear factors.

a. $4x^2 - 9$ **b.** $-x^2 + 2x - 1$ **c.** $2x^2 + 7x - 4$

d. $6x^2 - 13x + 6$ **e.** $8x^2 - 2x - 1$ **f.** $x^2 + \frac{5}{6}x + \frac{1}{6}$

2. When a seesaw is balanced, each person's weight varies inversely with the distance from the center support. Tony, who weighs 105 pounds sits at one end of a 10-foot-long seesaw. Where should his father, who weighs 190 pounds, sit to balance the seesaw.

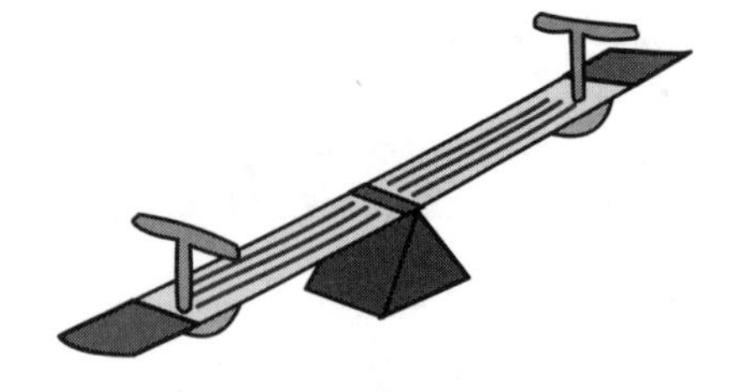

3. Graph the solution set to each system of inequalities.

a. $x \geq 4$
$y \geq -2$

b. $x - y < 3$
$2x + y > 6$

c. $x > 0$
$2y < 10$
$y > \frac{x}{3} + 2$

4. Consider the data in the table below.

Years of Schooling Beyond High School, *x*	1.5	2.7	5.6	1.2	4.2	1.9	2.5	2.9	4.7
Monthly Salary in Hundreds for First Job, *y*	3.8	9.2	22.2	10.4	23.9	13.6	8.3	10.1	26.1

a. Is this set of data reasonably modeled by a line? Explain.

b. The regression line has a slope of 4.44. Interpret this slope in the context of the data.

c. Samantha just completed a four-year college degree. Her sister, Mia, just completed a two-year associates' degree. Both are starting their first jobs. Use your model to predict the difference in their monthly salaries.

5. Write each of the following algebraic expressions in standard polynomial form.

a. $(2x^2 - 2x)(3x + 1)$ **b.** $(8 + 4x)^2$ **c.** $(2x - 3)(x + 4)$

d. $(4x + 1)(3x - 2)$ **e.** $\left(x - \frac{2}{3}\right)(3x + 6)$ **f.** $(4x^2 + 1)(2x - 1)$

Exercise Set 18

6. In the given figure, $\angle A \cong \angle EBC$, $AB = CD$, and $AF = BE$. Prove or disprove each statement.

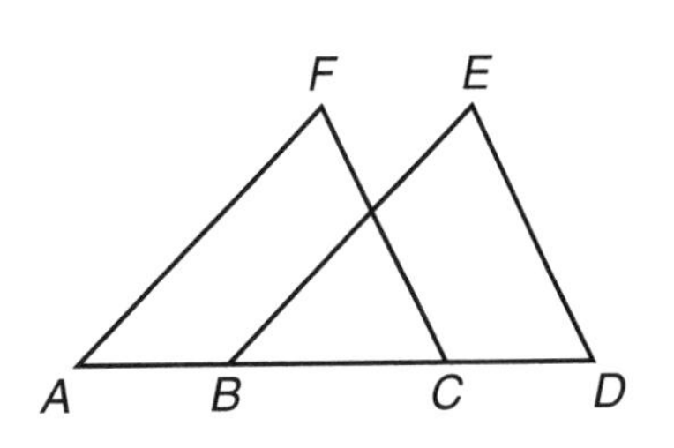

   a. $\overleftrightarrow{AF} \parallel \overleftrightarrow{BE}$

   b. $\triangle AFC \cong \triangle BED$

   c. $\angle F \cong \angle E$

7. Write each expression in an equivalent, simplified form.

   a. $\sqrt{49}\,\sqrt{25}$ **b.** $\sqrt{128}\,\sqrt{112}$ **c.** $\left(\frac{64}{121}\right)^{\frac{1}{2}}$ **d.** $3\sqrt{93}\,\sqrt{121}$

   e. $5\sqrt{18} \cdot 2\sqrt{33}$ **f.** $\left(\frac{8}{27}\right)^{\frac{1}{3}}$ **g.** $\left(\frac{38}{25}\right)^{\frac{1}{2}}$ **h.** $\frac{6\sqrt{18}}{\sqrt{6}}$

8. For each periodic function, find the coordinates (x, y) of the indicated points on its graph where $-2\pi \le x \le 2\pi$.

   a. $y = 5\sin x$; x-intercepts

   b. $y = \cos(x) + 4$; maximum points

   c. $y = 2\cos x$; x-intercepts

   d. $y = 2\sin(x) - 1$; minimum points

9. Find exact solutions to each quadratic equation.

   a. $x^2 - 8x = 16$ **b.** $x^2 + x - 11 = 0$

   c. $(x + 1)(x - 6) = 4$ **d.** $-2x - 63 + x^2 = 0$

   e. $2x^2 + 3x - 20 = 0$ **f.** $-x^2 - 10x = 24$

10. Consider the matrices $A = \begin{bmatrix} 4 & 1 \\ 2 & 3 \end{bmatrix}$, $B = \begin{bmatrix} 5 & -2 \\ 0 & 1 \end{bmatrix}$, $C = \begin{bmatrix} 1 & 0 \\ 0 & -1 \end{bmatrix}$, and $D = \begin{bmatrix} 4 \\ 3 \end{bmatrix}$.

    a. Find $A + B + C$.

    b. Find $A - B + C$.

    c. Find $C \times B$. Is $C \times B = B \times C$?

    d. Find $A \times B \times D$.

Exercise Set 19

1. The number of watts of power generated by a windmill varies directly with the cube of the wind speed in miles per hour. The constant of variation for a particular windmill is 0.012.

a. Write a modeling equation for this situation.

b. Describe the practical domain and range for this situation.

c. How much power is generated if the wind is blowing at 20 mph?

d. How fast must the wind be blowing in order to generate 200 watts of power?

e. If the wind speed doubles, what happens to the amount of power generated?

2. Solve each equation.

a. $3x + 4(x - 2) = 2(6 - x)$

b. $(x + 1)(x + 4) = -2$

c. $\frac{1}{4}(2x + 3) = \frac{1}{2}(5x + 9)$

d. $\frac{2}{x + 1} = \frac{5}{x}$

e. $6 \cdot 4^x = 96$

f. $\frac{-1}{3x^2} = -300$

3. What algebraic property or properties justify each step in this simplification of a polynomial expression?

a. $3(x^2 + 2x) - x(2 - 3x) + 5(x + 2) = 3x^2 + 6x - (2x - 3x^2) + 5x + 10$

b. $= 3x^2 + 6x - 2x + 3x^2 + 5x + 10$

c. $= 6x^2 + 9x + 10$

4. Write each expression in an equivalent, simplified form using only positive exponents. Assume all variables are nonzero.

a. $(2x^2)^3 - 9x^6$

b. $(y^{-1})^{-2}$

c. $\frac{x^2y^6}{3x^{-1}y^2}$

d. $(-4x^3)^3$

e. $6x^6 \cdot 2x^6$

f. $(8x^{-1})\left(\frac{1}{2}x^3y^{-2}\right)$

5. In quadrilateral $ABCD$ to the right, $\overline{AC}$ is a diagonal, $\overline{AB} \cong \overline{CD}$, $\angle BAC \cong \angle DCA$. Prove or disprove each statement.

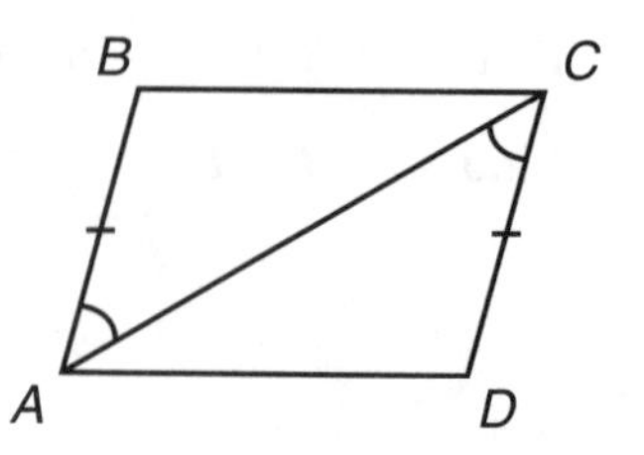

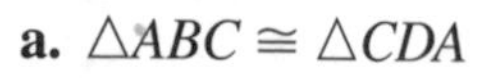

a. $\triangle ABC \cong \triangle CDA$

b. $\overline{BC} \parallel \overline{AD}$

c. $ABCD$ is a parallelogram.

6. Find exact solutions to each quadratic equation.

a. $x^2 - 11x + 20 = 0$

b. $x^2 - 2x - 10 = 0$

c. $x^2 + 2x - 35 = 0$

d. $-x^2 + 4x + 21 = 0$

7. Find the area of each triangle.

a.

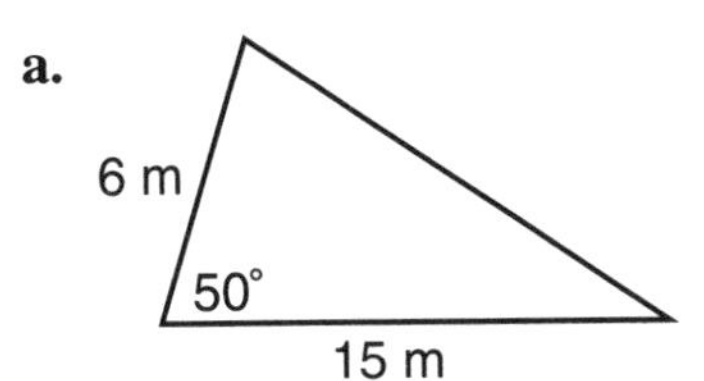

b.

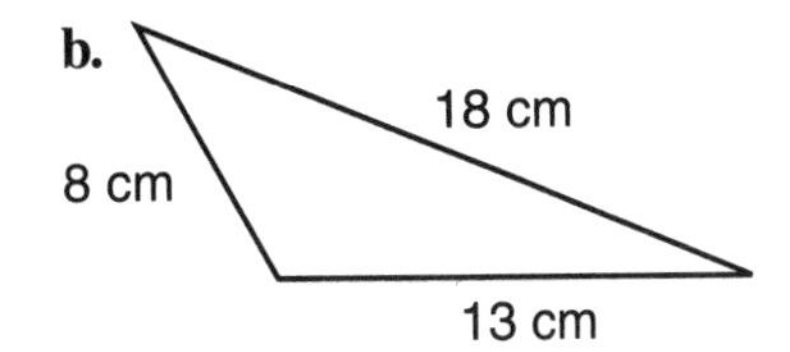

8. Approximately 30% of the students at a large state university are from North Carolina. Suppose you randomly stop students walking to the library until you find one who is from North Carolina.

a. What is the probability that the first person you stop is from North Carolina?

b. What is the probability that you must stop three students before identifying one from North Carolina?

c. Is having to stop six people before identifying a student from North Carolina a rare event?

9. $\triangle PQR$ has vertices $P(0, 0)$, $Q(6, 0)$, and $R(a, b)$.

a. Find values of a and b so that $\triangle PQR$ is a right triangle.

b. Find values of a and b so that $\triangle PQR$ is an isosceles triangle.

10. Consider the weighted graph below.

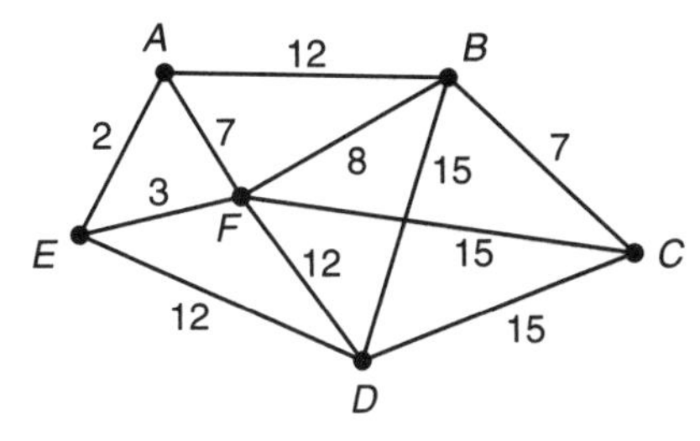

a. Find a minimal spanning tree.

b. What is the length of the minimal spanning tree?

c. Find a Hamiltonian circuit for the graph.

d. What is the length of your Hamiltonian circuit?

Exercise Set 20

1. If a basketball is properly inflated, it should rebound to about $\frac{1}{2}$ the height from which it is dropped. Ingrid initially drops a properly inflated basketball from a height of 64 inches.

a. What equation shows how to calculate the rebound height y as a function of the number of times x the basketball has bounced?

b. What equation using *NOW* and *NEXT* shows how the rebound height decreases with each bounce?

c. How many bounces have occurred if the rebound height is 8 inches?

d. Find the rebound height if four bounces have occurred.

e. Find the range of bounces for which the ball remains within 1 foot of the floor.

2. A triangular lot has sides of 215, 185, and 125 meters. Find the measures of the angles at the corners.

3. Write each combination of fractional expressions as a single fractional expression.

a. $\frac{3}{4} - \frac{5}{x}$ **b.** $\frac{x}{3} + \frac{2}{x}$ **c.** $\frac{4 + 3x}{x} - \frac{x}{3}$

d. $\frac{2}{x} + \frac{3}{4} - \frac{x}{2}$ **e.** $\frac{7 + x}{4} - \frac{3 + x}{5}$ **f.** $\frac{3}{x - 1} + \frac{2}{x + 1}$

4. In the figure at the right, $\angle B \cong \angle D$. Prove or disprove each statement.

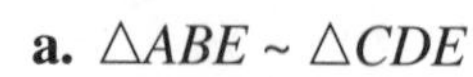

a. $\triangle ABE \sim \triangle CDE$

b. $\overleftrightarrow{AB} \parallel \overleftrightarrow{CD}$

c. If, in addition, $AE = 6$ and $EC = 3$, then $AB = 2DC$.

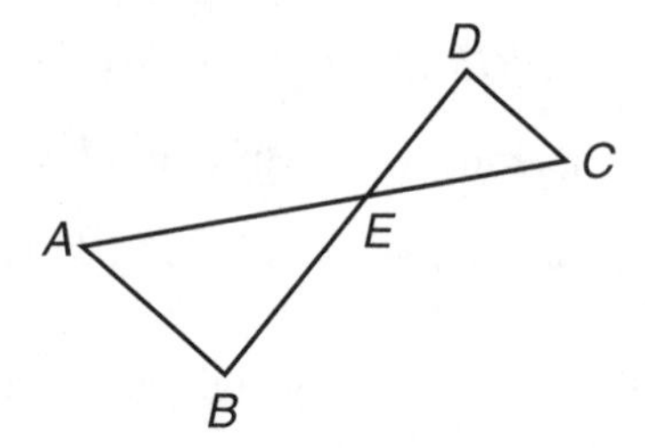

5. Solve each equation for x in terms of the other variables. Assume all variables are positive.

a. $x^2 + a = b$ **b.** $2x + 7 = bx + 9$ **c.** $\frac{c}{x + 4} = \frac{a}{b}$

d. $\sqrt{x + 2a} = b - 4$ **e.** $\frac{ax}{b} = \frac{c}{d}$ **f.** $ax^3 + b = c$

Exercise Set 20

6. A spinner A is divided into three equal sections numbered 1, 2, and 3. A second spinner B is divided into four equal sections numbered 2, 5, 6, and 8. Each pointer is spun. Find the probability of each event:

a. A 1 on spinner A and a 5 on spinner B

b. A 2 on spinner A and not a 2 on spinner B

c. An even number on spinner A and an even number on spinner B

d. The sum of the two numbers is less than 8.

7. Write each expression in standard polynomial form.

a. $(r + 4)(r - 3)$ **b.** $(2r + 5)(3r - 1)$ **c.** $(x - 3)(x - 4)$

d. $(y + 5)(3 - y)$ **e.** $(2x + 1)(x^2 - 3x + 4)$ **f.** $(3x - 2t)^2$

8. Solve each quadratic inequality.

a. $x^2 - x - 6 > 0$ **b.** $x^2 + 3x - 28 \leq 0$

c. $x^2 - 2x - 3 \geq 0$ **d.** $-x^2 - 8x - 15 < 0$

9. State the domain and range of each function.

a. $f(x) = 6 + 2.5x$ **b.** $f(x) = 3 \cdot 2^x$

c. $f(x) = \sqrt{x + 2}$ **d.** $g(x) = -x^2 + 3$

e. $h(x) = \dfrac{4}{x - 1}$ **f.** $g(x) = 4\sin x - 1$

10. A baseball park is laid out as in the diagram below.

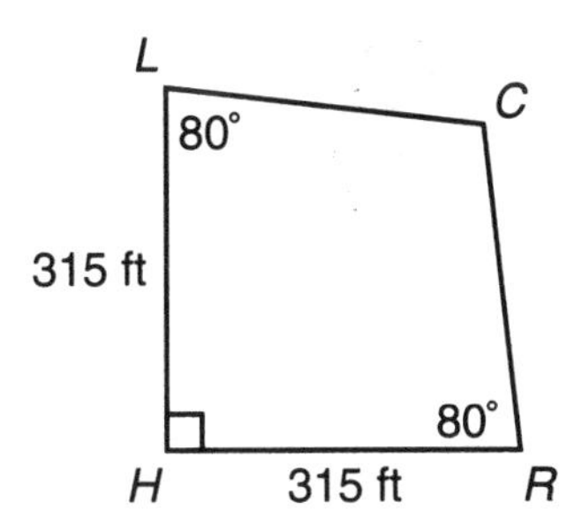

a. Find the distance from H to C.

b. How far is point C from point R in the right field corner?

Exercise Set 2B

6. A spinner A is divided into three equal sections numbered 1, 2, and [illegible]. A second spinner B is divided into four equal sections numbered 1, 2, 5, [illegible], and [illegible]. Each pointer is spun. Find the probability of each event.

   a. [illegible] on spinner A and a 5 on spinner B

   b. [illegible] 2 on spinner A and not a 2 on spinner B

   c. [illegible] number on spinner A and an [illegible] number on spinner B

   d. The sum of the two numbers is [illegible]

7. Write each expression as a [illegible] rational [illegible].

   a. [illegible]

   b. [illegible]

8. Solve each quadratic inequality.

   a. [illegible]

   b. x^2 [illegible] $28 < 0$

   c. [illegible] 2 [illegible] 0

   d. $-x^2$ [illegible]

9. State the domain and range of each function.

   a. [illegible]

   b. [illegible]

   c. [illegible] $\sqrt{x+5}$

   d. [illegible] 3

   e. [illegible]

   f. $g(x) = 4\sin x$

10. A baseball diamond is shown in the diagram below.

   a. Find the distance from H to C.

   b. How far is point C from point K on the right field corner?

Practicing for Standardized Tests

As you work closely with your classmates and teachers on a daily basis, they will have a good idea of what you know and are able to do with respect to the mathematics you are studying this year. However, your school district or state department of education may ask you to take tests that they design to measure the achievement of all students, classes, or schools in the district or state. Colleges also use external standardized tests, like the ACT and SAT, to compare the knowledge of different students applying for admission or scholarships.

External standardized tests usually present assessment tasks in formats that can easily be scored to produce simple percent-correct ratings of your knowledge. If you want to perform well on such standardized tests, it helps to have some practice with test items in multiple-choice formats. The following ten sets of multiple-choice tasks have been designed to give you that kind of practice and to offer some strategic advice in working on such items.

Summarized below are test-taking strategies developed in the Course 2 RAP book.

- Work backwards from choices.
- If a diagram is not provided for a geometry problem, draw and label one.
- Replace variables with numbers.
- Break complex geometric shapes into simpler shapes if a particular formula cannot be remembered.
- Memorize important facts and formulas.
- Answer the easy questions first; then answer the more difficult ones.

- Use the table-building capability of your calculator to aid in reasoning with complicated function rules.
- Know the Pythagorean Theorem and how to use it.
- Be careful in applying proportional reasoning to linear relationships.
- For general problem situations, create and analyze a specific example.

Additional *Test Taking Tips* may be found at the end of each of the practice sets.

Practice Set 1

Solve each problem. Then record the letter that corresponds to the correct answer.

1. If the length of a leg of a 45° −45°−90° triangle is 5 cm, how many centimeters long is the hypotenuse?

(a) 5 **(b)** $5\sqrt{2}$ **(c)** $5\sqrt{3}$ **(d)** 10 **(e)** 25

2. Which one of the following equations matches the graph below?

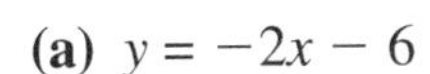

(a) $y = -2x - 6$

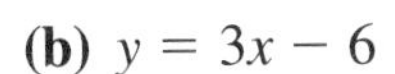

(b) $y = 3x - 6$

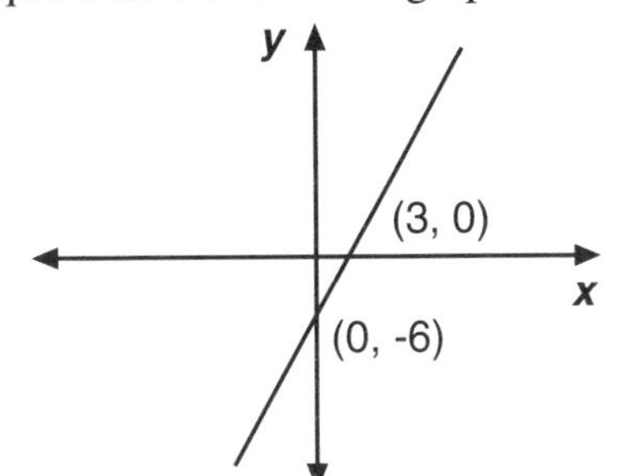

(c) $y = -6x - 2$

(d) $y = -2x - 3$

(e) $y = 2x - 6$

3. $|-9| - 3|-2| + |-6| =$

(a) −21 **(b)** −9 **(c)** −6

(d) 9 **(e)** 21

4. Given the lengths shown, in inches, find the length, in inches, of the third side of the triangle.

(a) $2\sqrt{5}$

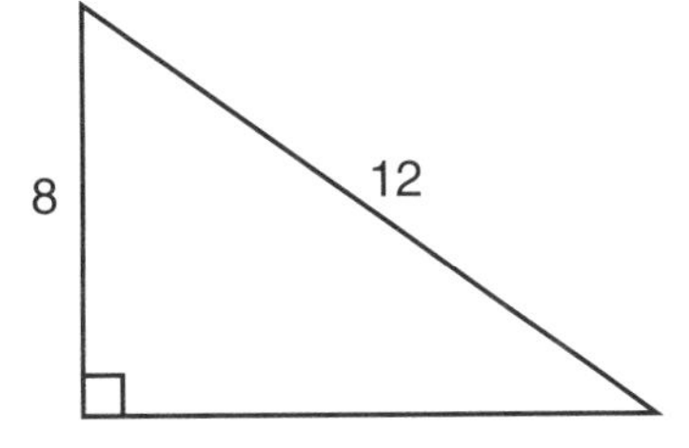

(b) 4

(c) $4\sqrt{5}$

(d) $8\sqrt{5}$

(e) $4\sqrt{13}$

5. Multiply $x^2y^3z^5 \cdot 2xy^2z^5$.

(a) $x^4y^7z^{15}$

(b) $2x^2y^5z^{10}$

(c) $2x^3y^5z^{10}$

(d) $2x^3y^6z^{10}$

(e) $2x^2y^6z^{25}$

6. Which of the following expressions is *not* equivalent to $4(7 + x)$?

(a) $(7 + x)4$

(b) $4(x + 7)$

(c) $(x + 7)4$

(d) $7(4 + x)$

(e) $4x + 28$

7. If $x + y = 9$ and $y - x = 7$, then $x^2 + y^2 =$

(a) 1 **(b)** 2 **(c)** 8 **(d)** 63 **(e)** 65

8. If $\begin{bmatrix} 2 & 0 \\ x & 3 \end{bmatrix} \begin{bmatrix} 0 & 4 \\ 2 & 1 \end{bmatrix} = \begin{bmatrix} 0 & 8 \\ 6 & 9 \end{bmatrix}$, then $x =$

(a) $\frac{2}{3}$

(b) $\frac{3}{2}$

(c) 3

(d) 4

(e) 6

9. Which of the following is a good first step in solving $(x + 3)(x - 2) = 14$?

(a) Set the sum of the factors equal to 14.

(b) Set $(x + 3)$ equal to 2, and set $(x - 2)$ equal to 7.

(c) Set each factor equal to 0.

(d) Set each factor equal to 14.

(e) Multiply to remove the parentheses.

10. In the figure, $\overleftrightarrow{AD} \parallel \overleftrightarrow{HE}$ and $\overleftrightarrow{BH} \parallel \overleftrightarrow{DF}$. Which one of the following quadrilaterals is a parallelogram?

(a) *ADEH*

(b) *CDFG*

(c) *BDEH*

(d) *BDFH*

(e) *CDEG*

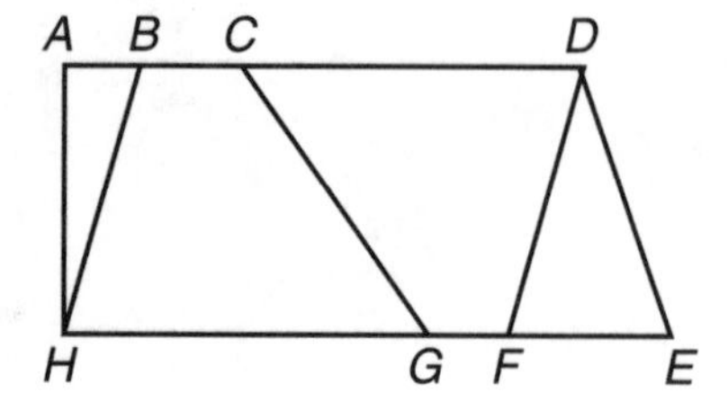

Practice Set 1

Test Taking Tip

Use your calculator to evaluate expressions.

When finding numerical answers, the form of your answer may not match any of the choices given. Use your calculator to find a decimal approximation of your answer. Then use your calculator to find the expression that has the same approximation from among the given choices.

Example Look back at Item 4 on page 89. To use this strategy, first use the Pythagorean Theorem to determine that the length of the third side is $\sqrt{144 - 64} = \sqrt{80}$. The decimal approximation of this is 8.94427.

For choice (a): $2\sqrt{5} \approx 4.472136$.

For choice (c): $4\sqrt{5} \approx 8.94427$.

So, the answer is (c).

- Find, if possible, another test item in the practice set for which this strategy might be helpful. Try it.
- Keep this strategy in mind as you work on future practice sets.

Practice Set 2

Solve each problem. Then record the letter that corresponds to the correct answer.

1. The rectangle below has lengths as marked, in units. What is the area, in square units, of the shaded triangle if it is enclosed in the rectangle as shown?

(a) 25

(b) 34

(c) 50

(d) 68

(e) 136

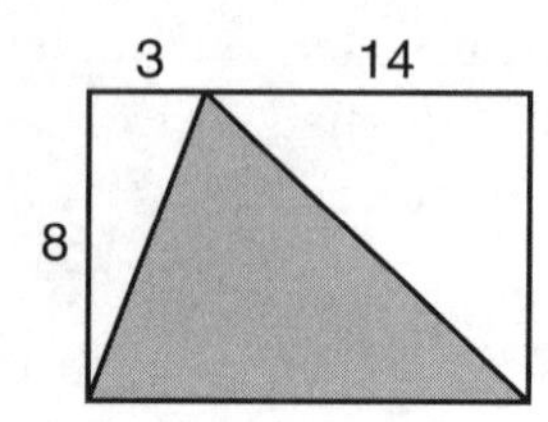

2. Of a sample of 100 Bedford High School juniors who were asked the questions:

1. Are you planning to go to college?

2. Are you enrolled in a math class?

Twelve students answered "no" to both questions, 75 answered "yes" to Question 1, and 80 answered "yes" to Question 2. How many answered "yes" to both questions?

(a) 5

(b) 43

(c) 55

(d) 67

(e) 88

3. If the measure of $\angle MRY = 65^\circ$ and the measure of $\angle AMR = 40^\circ$, find the measure of $\angle MAR$.

(a) 25°

(b) 40°

(c) 55°

(d) 105°

(e) 115°

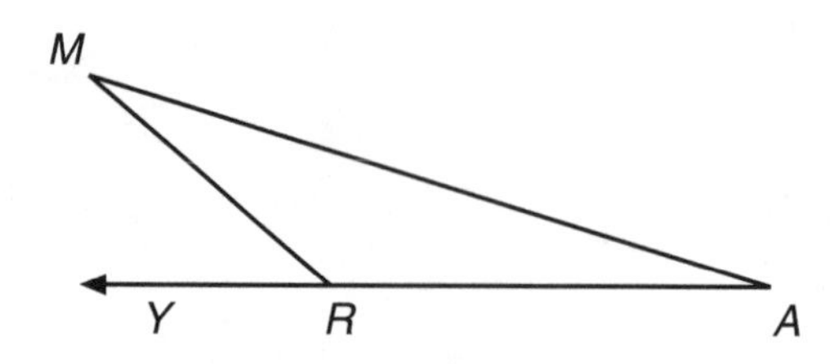

4. Given a bag with 8 red candies, 5 yellow candies, 3 blue candies, and 6 green candies, find the probability of drawing a red candy from the bag on the first draw.

(a) $\frac{1}{8}$

(b) $\frac{4}{11}$

(c) $\frac{8}{21}$

(d) $\frac{4}{7}$

(e) $\frac{7}{11}$

5. If x is 50 percent of y and y is 50 percent of z, then z is what percent of x?

(a) 25%

(b) 50%

(c) 100%

(d) 125%

(e) 400%

6. In a group of 200 students, more students are taking Spanish than are taking German. If 120 students are taking Spanish and 40 are taking neither Spanish nor German, what is the maximum number of students who could be taking both languages?

(a) 39

(b) 59

(c) 79

(d) 119

(e) 159

7. $\triangle PQR$ has vertices $P(-3, 2)$, $Q(-1, 5)$, and $R(3, 2)$. What is the area of $\triangle PQR$ in square units?

(a) 6

(b) 9

(c) 12

(d) 15

(e) 18

8. If $x < 0$, which of the following must be true?

(a) $x - 3 < 3x$

(b) $x - 3 < 3 - x$

(c) $-3x < x^2$

(d) $x^3 > x + 3$

(e) $x - 3 > x + 3$

9. $\boxed{a \mid b \mid c \mid d \mid e}$ is a *mean number strip* if b is the arithmetic mean of a and c, c is the mean of b and d, and d is the mean of c and e. If $\boxed{2 \mid 5 \mid c \mid d \mid e}$ is a mean number strip, what is the value of e?

(a) 8

(b) 12

(c) 14

(d) 16

(e) 20

10. What is the value of $f(-2)$ when $f(x) = (3x - 1)(x^2 - 10x + 6)$?

(a) -210

(b) -154

(c) -150

(d) -35

(e) -17

Practice Set 2

Test Taking Tip

Use Venn diagrams to aid in reasoning about counting problems.

Example Look back at Item 2 on page 92. To use this strategy, draw and label a Venn diagram as shown below.

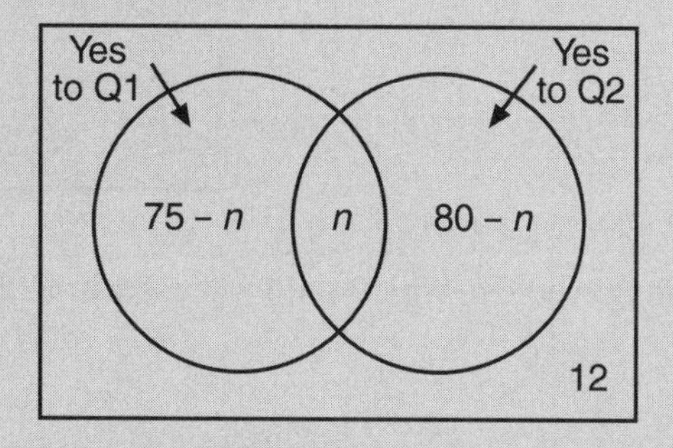

The number of students n that answered "yes" to both questions is represented by the region where the two circles overlap. Then the number of students who answered "yes" to only Question 1 is $75 - n$, and the number who answered "yes" to only Question 2 is $80 - n$. Since 12 students answered "no" to both questions, it must be the case that:

$$100 = (75 - n) + n + (80 - n) + 12 = 167 - n.$$

So, $n = 67$.

The correct answer is (d).

- Find, if possible, another test item in the practice set for which this strategy might be helpful. Try it.
- Keep this strategy in mind as you work on future practice sets.

Practice Set 3

Solve each problem. Then record the letter that corresponds to the correct answer.

1. A semicircle is joined to a square as shown in the figure below, with lengths given in units. What is the area, in square units, of the figure?

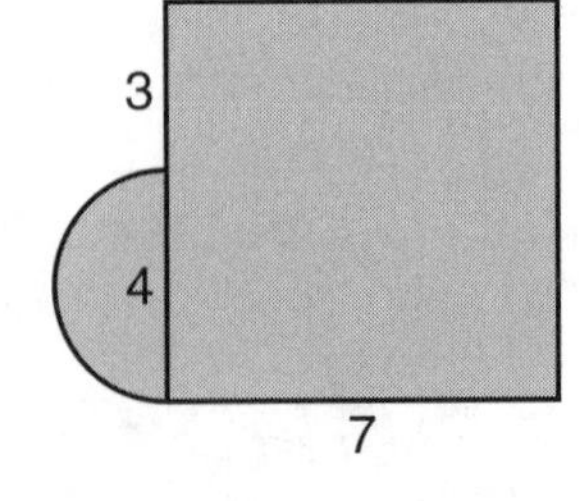

(a) $28 + 4\pi$

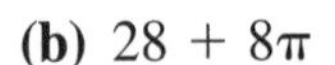
(b) $28 + 8\pi$

(c) $49 + 2\pi$

(d) $49 + 4\pi$

(e) $49 + 16\pi$

2. In the right triangle below, which ratio represents the tangent of $\angle C$?

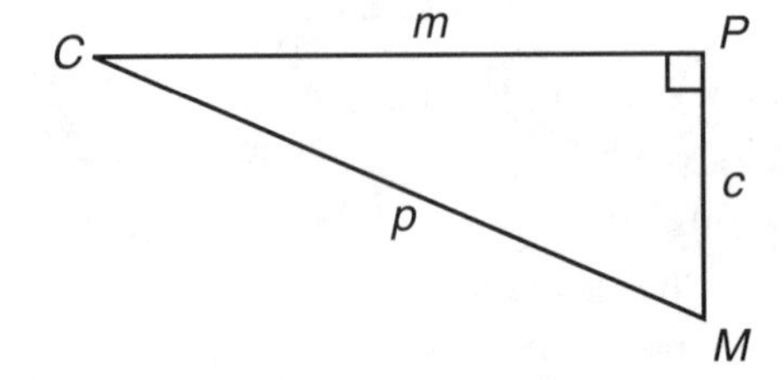

(a) $\frac{c}{m}$

(b) $\frac{m}{c}$

(c) $\frac{c}{p}$

(d) $\frac{m}{p}$

(e) $\frac{p}{c}$

3. In a rectangle, the sides have measurements of $2x + 1$ and $3x - 5$ units. Find an expression for the area of the rectangle, in square units.

(a) $5x + 4$

(b) $5x - 5$

(c) $6x^2 - 5$

(d) $6x^2 - 7x - 5$

(e) $6x^2 - 13x - 5$

4. A person has to be at least 4.5 feet tall to ride the Raptor at Cedar Point. If t represents height in inches, which of the following inequalities represents this situation?

(a) $t \geq 4.5$

(b) $t \leq 4.5$

(c) $t < 54$

(d) $t > 54$

(e) $t \geq 54$

5. If, for any number, k, $k* = k + 2$ and $*k = k - 2$, which of the following is *not* equal to (*6)(4*)?

(a) (*10)(1*)

(b) (10*)(*4)

(c) *14 + 10*

(d) 2(10*)

(e) (*4)(12*)

6. Find the expression that represents all solutions for $|2x - 1| < 3$.

(a) $x < 2$

(b) $2 < x < -1$

(c) $-1 < x < 2$

(d) $x > 0$

(e) $x > 2$

7. If $x > 0$ and $x \cdot x \cdot x \cdot x = x + x + x + x$, what is the value of x?

(a) $\frac{1}{4}$

(b) 1

(c) $\sqrt{2}$

(d) $\sqrt[3]{4}$

(e) 4

8. If $4x - 12 = 10$, what is the value of $x - 3$?

(a) 2.5

(b) 3.3

(c) 4.75

(d) 5.5

(e) 6

9. If $4x - 3y = 11$ and $-5x + 2y = -12$, then the solution (x, y) is:

(a) $(-2, 11)$

(b) $(2, -1)$

(c) $\left(\frac{14}{5}, 1\right)$

(d) $\left(3, \frac{1}{3}\right)$

(e) $(6, -1)$

10. In factored form, the quadratic equation $2x^2 + kx + 12 = 0$ can be written as $(2x - 3)(x - 4) = 0$. What must be the value of k?

(a) -11

(b) -8

(c) -5

(d) -3

(e) 11

Practice Set 3

Test Taking Tip

Use graphs to solve equations and inequalities.

If you aren't sure how to solve an equation or inequality by reasoning with the symbols themselves, you may still be able to estimate the solution with a graph.

Example Look back at Item 6 on page 97. To use this strategy, put the expression on each side of the inequality into the function menu. Then look at the graphs.

$y_1 = \text{abs}(2x - 1)$

$y_2 = 3$

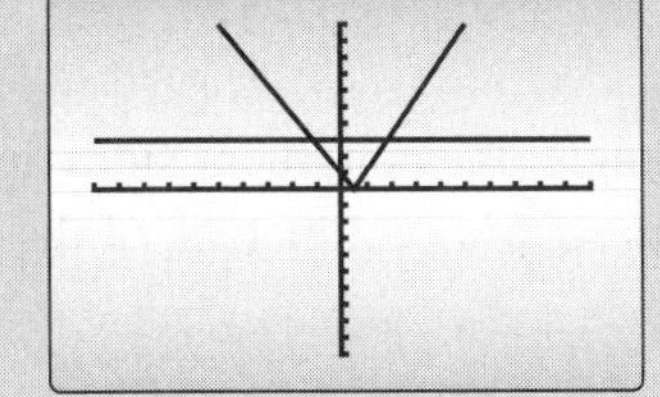

From the graph, you can see that the absolute value graph is below the graph of $y = 3$ for x-values between -1 and 2. So, the answer is (c).

- Find, if possible, another test item in the practice set for which this strategy might be helpful. Try it.
- Keep this strategy in mind as you work on future practice sets.

Practice Set 4

Solve each problem. Then record the letter that corresponds to the correct answer.

1. If the radius of a circle is 10 units, what is the measure of the circumference in units?

(a) 10π

(b) 20π

(c) $10\pi^2$

(d) 100π

(e) $20\pi^2$

2. In an isosceles triangle, the congruent sides are each 5 cm long. The base is 8 cm. What is the area, in square centimeters, of the triangle?

(a) 12

(b) 18

(c) 20

(d) 24

(e) 40

3. If $m = 2$, $a = 7$, and $r = 4$, what is the value of $am^2 + \frac{r^3}{2}$?

(a) 36

(b) 43

(c) 60

(d) 153

(e) 228

4. Choose the correct equation for the graph given.

(a) $y = |x + 5|$

(b) $y = |x| + 5$

(c) $y = |x - 5|$

(d) $y = |x|$

(e) $y = |x| - 5$

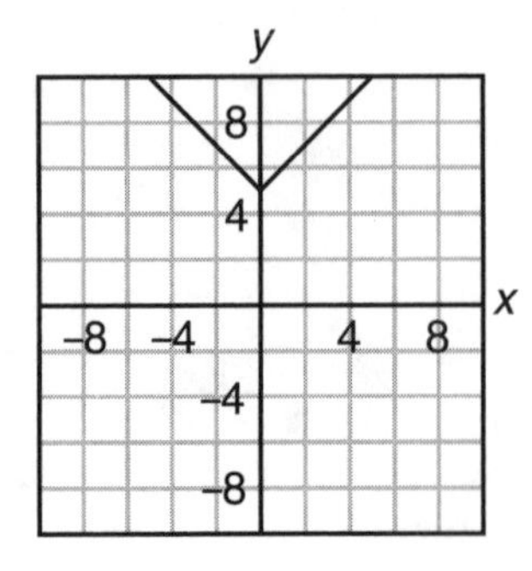

5. Suppose you roll two dice until you get a sum of 11. The probability you will get a sum of 11 for the first time on the third roll is

(a) $\frac{17}{5,832}$

(b) $\frac{289}{5,832}$

(c) $\frac{35}{46,656}$

(d) $\frac{1,225}{46,656}$

(e) $\frac{1}{18}$

6. Using the triangle at the right, the sine of the smallest angle is:

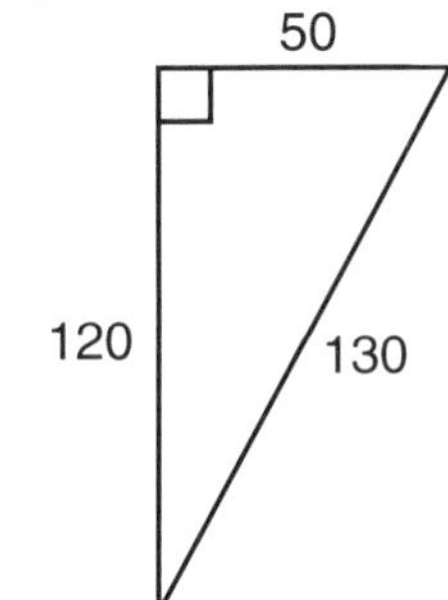

(a) $\frac{5}{12}$

(b) $\frac{5}{13}$

(c) $\frac{12}{13}$

(d) $\frac{12}{5}$

(e) $\frac{13}{5}$

7. If $x + y = 5$ and $x^2 + 3xy + 2y^2 = 40$, what is $x + 2y$?

(a) 5

(b) 6

(c) 7

(d) 8

(e) 9

8. If $\frac{3x}{7} = 9$, what is the value of $\frac{x}{3}$?

(a) $\frac{7}{9}$

(b) $\frac{9}{7}$

(c) 7

(d) 21

(e) 63

9. In right triangle ABC below, the cosine of $\angle A$ is $\frac{3}{7}$. What is the sine of $\angle A$?

(a) $2\sqrt{10}$

(b) $\frac{2\sqrt{10}}{3}$

(c) $\frac{2\sqrt{10}}{7}$

(d) $\frac{3\sqrt{40}}{40}$

(e) $\frac{7\sqrt{40}}{40}$

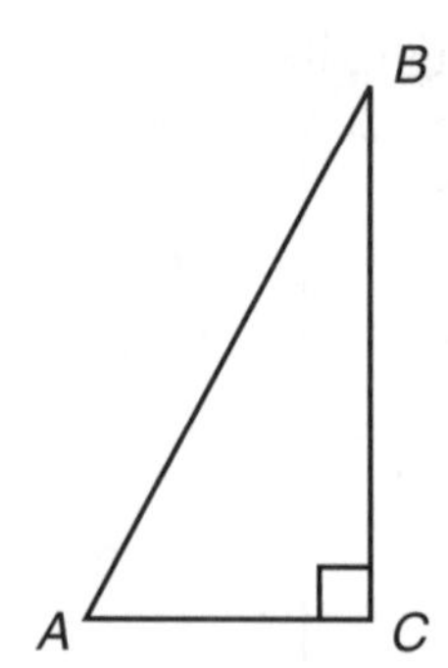

10. What is the slope of a line that is perpendicular to the line with equation $2y + 7x = 14$?

(a) $-\frac{7}{2}$

(b) $\frac{1}{7}$

(c) $\frac{2}{7}$

(d) $\frac{7}{2}$

(e) 7

Practice Set 4

Test Taking Tip

Know and be able to use the definitions of the sine, cosine, and tangent ratios.

Questions about right triangles often involve sines, cosines, and tangents in some way. Sometimes you are asked to calculate one of the trigonometric ratios; at other times, you need to observe that the calculation of a trigonometric ratio is necessary. Your calculator has built-in trigonometric functions that can be used to obtain approximate values.

Example Look back at Item 9 on page 102. The expression for sine of angle A requires the length of side BC since $\sin A = \frac{\text{side opposite}}{\text{hypotenuse}}$. $BC^2 = 7^2 - 3^2$ by the Pythagorean Theorem. Thus, $BC = \sqrt{40}$ and $\sin A = \frac{\sqrt{40}}{7} = \frac{2\sqrt{10}}{7}$. So, the answer is (c).

- Find, if possible, another test item in the practice set for which this strategy might be helpful. Try it.
- Keep this strategy in mind as you work on future practice sets.

Practice Set 5

Solve each problem. Then record the letter that corresponds to the correct answer.

1. If $a > 0$ and $b > 0$ and the slope of the line passing through $(-2a, a)$ and $(b, 3a)$ is 2, which of the following is an expression for b in terms of a?

(a) $-a$

(b) $-2a$

(c) $-2a - 2$

(d) $3a$

(e) $4a$

2. Olivia is planning the Mathematics Club annual picnic. She may choose from 3 meat selections, 4 types of vegetables, and 5 kinds of cookies. How many different menus, each with a meat, a vegetable, and a cookie, could she select?

(a) 12

(b) 17

(c) 23

(d) 35

(e) 60

3. In the figure shown, determine the value of x.

(a) 10°

(b) 30°

(c) 50°

(d) 70°

(e) 80°

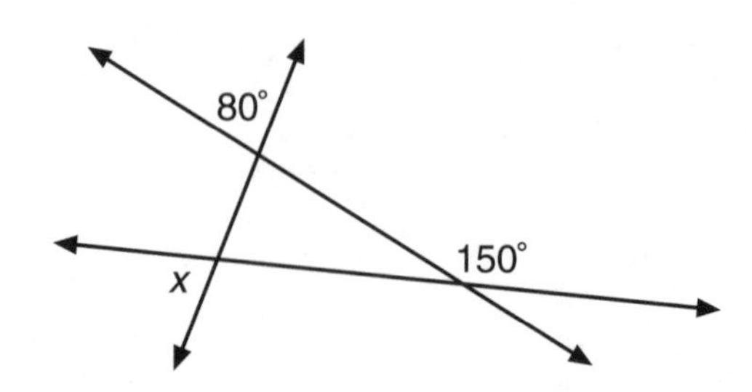

4. A school's honor society has 150 members: 70 boys and 80 girls, of whom 40 are juniors and 110 are seniors. What is the smallest possible number of senior girls in the society?

(a) 0

(b) 20

(c) 40

(d) 60

(e) 80

5. If $3x^2 + 10x - 8 = 0$, what are the two solutions?

(a) $x = -2$ and $x = -4$

(b) $x = 2$ and $x = 4$

(c) $x = \frac{2}{3}$ and $x = 4$

(d) $x = \frac{2}{3}$ and $x = -4$

(e) $x = -\frac{2}{3}$ and $x = -4$

6. If $p + q = 5$ and $2pq = 8$, what is $p^2 + q^2$?

(a) 1

(b) 3

(c) 9

(d) 13

(e) 17

7. If $\sqrt{k}$ is an integer, which of the following must be integers?

I. $\sqrt{\frac{k}{4}}$ II. $\left(\sqrt{5k}\right)^2$ III. $\sqrt{9k}$

(a) None

(b) I and II only

(c) I and III only

(d) II and III only

(e) I, II, and III

8. If $m^2 - 4n^2 = 32$ and $m + 2n = 8$, what is the value of $m - 2n$?

(a) -4

(b) $\frac{1}{4}$

(c) 2

(d) 4

(e) 16

9. The lengths of the sides of a triangle are 5 meters, 10 meters, and 13 meters, as depicted in the diagram. What is the cosine of the smallest of the triangle's three angles?

(a) $\frac{-244}{260}$

(b) $\frac{-10}{13}$

(c) $\frac{5}{13}$

(d) $\frac{10}{13}$

(e) $\frac{244}{260}$

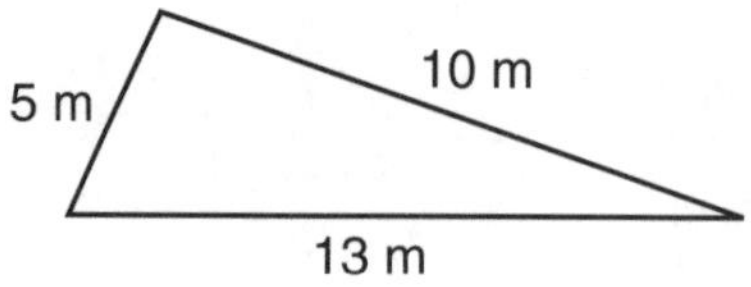

10. Approximately 25% of the cars sold are red. Michael is observing cars as they drive by his house. What is the probability that he has to watch at least three cars go by in order to see a red one?

(a) 0.140625

(b) 0.25

(c) 0.4375

(d) 0.5625

(e) 0.75

Practice Set 5

Test Taking Tip

Factoring a quadratic expression often makes a question transparent.

College entrance examination authors think it is important for students to be able to factor several simple quadratic expressions such as $ax^2 \pm bx$, $x^2 + 2ax + a^2$, and $x^2 - y^2$. However, you are seldom asked directly to factor. Rather you are asked a question that is simple once you see that factoring a quadratic expression is involved.

Example Look back at Item 8 on page 106. To use factoring, notice that in $m^2 - 4n^2 = 32$, the lefthand expression is the difference of two squares, m^2 and $(2n)^2$. Thus, $m^2 - 4n^2 = (m + 2n)(m - 2n) = 32$. Since $m + 2n = 8$, $m - 2n = 4$. So the answer is (d).

- Find, if possible, another test item in the practice set for which this strategy might be helpful. Try it.
- Keep this strategy in mind as you work on future practice sets.

Practice Set 6

Solve each problem. Then record the letter that corresponds to the correct answer.

1. A triangle has side lengths of 9 inches, 12 inches, and 15 inches. The perimeter of a square is 24 inches. By what amount, in square inches, do the areas of these two figures differ?

(a) 12

(b) 18

(c) 30

(d) 31.5

(e) 54

2. In the figure below, $\triangle KAL$ and $\triangle KZA$ are right triangles with lengths in units as marked. How many units long is AZ?

(a) 6

(b) $6\sqrt{3}$

(c) $12\sqrt{3}$

(d) 18

(e) 24

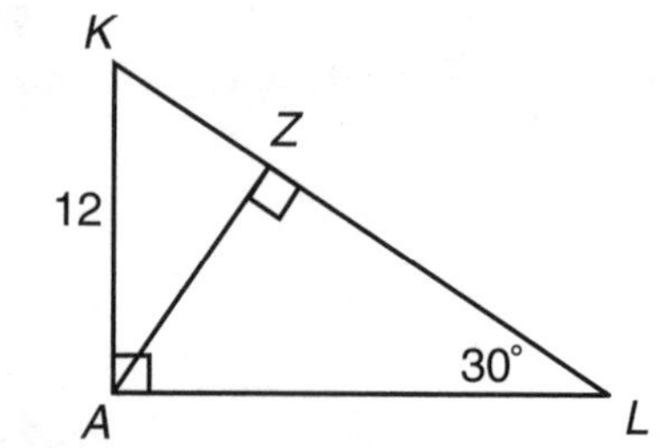

3. Which expression is equivalent to $6c(2c+4) + 5(c - 1)$?

(a) $40c$

(b) $12c^2 + 24c$

(c) $12c^2 + 28c$

(d) $12c^2 + 29c - 5$

(e) $36c^2 + 5c - 1$

4. If Joe's first four math test scores in the semester are 81, 93, 85, and 72, what score does he need on the fifth test to have an average of 85?

(a) 83

(b) 85

(c) 88

(d) 91

(e) 94

5. In a class of 100 students, 58 have their own computers, and 77 have their own calculators. If 18 have neither a computer nor a calculator, how many students have both?

(a) 5

(b) 19

(c) 24

(d) 53

(e) 58

6. The hypotenuse of a right isosceles triangle is 14 inches. Find the length, in inches, of a leg.

(a) 7

(b) $7\sqrt{2}$

(c) $7\sqrt{3}$

(d) 4

(e) $14\sqrt{2}$

7. If $x > 1$, which of the following decreases as x increases?

I. $1 - \sqrt{x}$ II. $\frac{x}{x + 1}$ III. $\frac{1}{1 - x}$

(a) NONE

(b) I only

(c) II only

(d) I and III only

(e) II and III only

8. The matrix below gives the dollar cost (in thousands) to build enclosed skywalks between several buildings in a downtown area. What is the minimum cost (in thousands of dollars) of building skywalks in order that people can walk from any building to any other using the skywalks?

(a) 7

(b) 8

(c) 9

(d) 10

(e) 11

$$\begin{array}{c} \\ A \\ B \\ C \\ D \\ E \end{array} \begin{array}{c} \begin{array}{ccccc} A & B & C & D & E \end{array} \\ \begin{bmatrix} 0 & 4 & 2 & 1 & 10 \\ 4 & 0 & 4 & 5 & 3 \\ 2 & 4 & 0 & 1 & 2 \\ 1 & 5 & 1 & 0 & 4 \\ 10 & 3 & 2 & 4 & 0 \end{bmatrix} \end{array}$$

9. The operation $*$ is defined as follows: For any positive numbers a and b, $a * b = \sqrt{a + 2\sqrt{b}}$. Which of the following is an integer?

(a) $4 * 9$

(b) $5 * 4$

(c) $7 * 16$

(d) $16 * 9$

(e) $9 * 25$

10. Which of the following describes all the values of x that satisfy $5 - 2(3 - x) \geq 5x - 3(x - 1)$?

(a) All real numbers

(b) $x < -1$

(c) $x \geq \frac{1}{2}$

(d) $x \leq \frac{1}{2}$

(e) No real numbers

Practice Set 6

Test Taking Tip

Memorize the relationships among the sides of 45°–45°–90° triangles and 30°–60°–90° triangles.

Questions on standardized tests often involve special right triangles: 45°–45°–90° and 30°–60°–90°. You should memorize the relationships among the sides of each type of triangle as indicated in the diagram below.

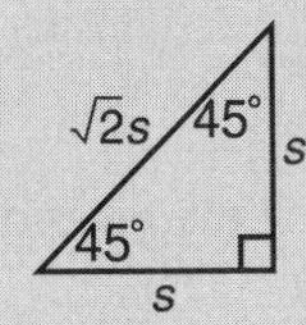

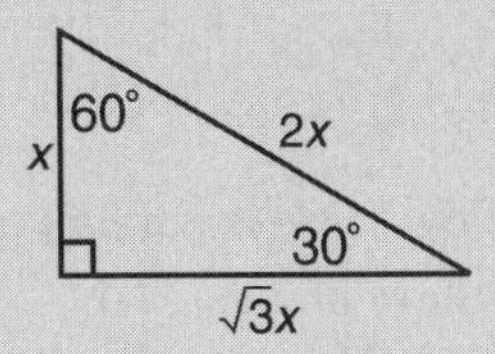

Example Look back at Item 2 on page 108. Since $\triangle KAL$ is 30°–60°–90°, so is $\triangle KAZ$. Since $KA = 12$ or $6 \cdot 2$, $AZ = 6 \cdot \sqrt{3}$. So the answer is (b).

- Find, if possible, another test item in the practice set for which this strategy might be helpful. Try it.
- Keep this strategy in mind as you work on future practice sets.

Practice Set 7

Solve each problem. Then record the letter that corresponds to the correct answer.

1. The area of a circle is 81π square inches. What is its circumference in inches?

(a) 9

(b) 4.5π

(c) 6π

(d) 9π

(e) 18π

2. In the figure below, $\overleftrightarrow{AM}$ is parallel to $\overleftrightarrow{RU}$. Use the measurements, given in inches, to find the length, in inches, of $\overline{TU}$.

(a) 3

(b) $\frac{10}{3}$

(c) 4

(d) $\frac{11}{2}$

(e) $\frac{15}{2}$

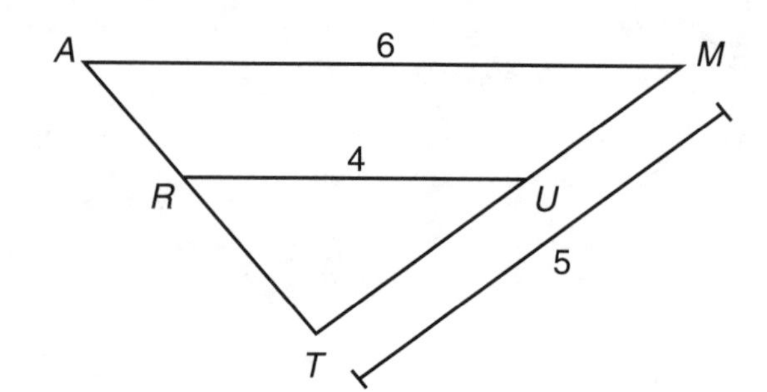

3. To play one of the games at a school carnival, you are blindfolded and then you draw one of 50 plastic eggs out of a basket. The color of the egg determines how much you win or lose. There are 15 red eggs for which you win 50¢. There are 10 blue eggs for which you win \$1.00. There are 25 yellow eggs for which you *pay* 25¢. What are the expected winnings for this game?

(a) 0.094¢

(b) 22.5¢

(c) 35¢

(d) 47.5¢

(e) \$11.25

4. In one month, 1,560 people purchased action and/or comedy videos at a discount store. 807 people purchased both action and comedy videos, 430 people purchased just action videos. How many people purchased just comedy videos?

(a) 323

(b) 377

(c) 753

(d) 1,130

(e) Impossible to determine from the information given

5. $\dfrac{56c^5d^2e}{-14c^7de^3} =$

(a) $-4c^2de^2$

(b) $\dfrac{-4d}{c^2e^2}$

(c) $\dfrac{-4c^2e^2}{d}$

(d) $\dfrac{-4}{c^2d^2e}$

(e) $42c^2de^2$

6. On the grid, which of the following points is the same distance from $P(1, 4)$ as it is from $Q(5, 2)$?

(a) (2, 2)

(b) (5, 5)

(c) (2, 1)

(d) (1, 2)

(e) (3, 4)

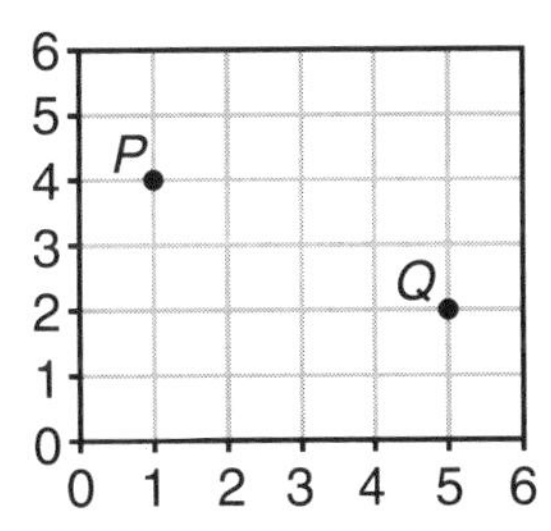

7. If, for any number b, $b\# = 2b$ and $\#b = \frac{b}{2}$, which of the following is *not* equal to (#18)(2#)?

(a) (9#)(#4)

(b) (4#)(#9)

(c) $\frac{9\#}{\#1}$

(d) (9#)(4#)

(e) $\frac{18\#}{\#2}$

8. If a, b, and c are nonzero numbers, and $6a = 8b$ and $4b = 9c$, then what is $\frac{a}{c}$?

(a) $\frac{1}{3}$

(b) $\frac{1}{2}$

(c) $\frac{3}{4}$

(d) $\frac{3}{2}$

(e) 3

9. If the endpoints of a diameter of a circle are $A(2, 10)$ and $B(-4, 2)$, where is the center of the circle?

(a) (3, 4)

(b) $(-1, 4)$

(c) (3, 6)

(d) $(-2, 5)$

(e) $(-1, 6)$

10. How many different numbers are solutions for the equation $5x + 17 = (x + 4)(x + 5)$?

(a) 0

(b) 1

(c) 2

(d) 3

(e) Infinitely many

Practice Set 7

Test Taking Tip

Eliminate options that are obviously incorrect; concentrate on the viable options.

Test makers often want you to determine one expression that is equivalent to a given one that may be rather complex. It takes time to do all the calculations, so you should seek clues that eliminate several of the options. To do this, look at the complex expression to determine its general characteristics (Is it quadratic? Is it cubic?), and then use these characteristics to eliminate options and, perhaps, to identify the correct option.

Example Look back at Item 10 on page 114. The given expression, when all like terms are combined, is a quadratic. You know a quadratic can have no more than two distinct solutions, so options (d) and (e) are eliminated. Since the condensed form, $x^2 + 4x + 3$, is not a perfect square, there is no duplication of roots. So there are exactly two different solutions. Option (c) is correct.

- Look back at Practice Sets 1 through 7 and find other test items for which eliminating obviously incorrect options might be helpful.
- Keep this strategy in mind as you work on future practice sets.

Practice Set 8

Solve each problem. Then record the letter that corresponds to the correct answer.

1. Two triangles are similar. The lengths of the sides of one triangle are 4, 9, and 11 units. The smallest side of the other triangle is 6 units. What is the perimeter, in units, of the larger triangle?

(a) 24

(b) 26

(c) 30

(d) 36

(e) 46

2. Given that the longer leg of a $30^\circ-60^\circ-90^\circ$ triangle measures 6 units, find the length, in units, of the hypotenuse.

(a) 3

(b) $2\sqrt{3}$

(c) $4\sqrt{3}$

(d) $6\sqrt{3}$

(e) 12

3. The radius of a circle is decreased by 20%. By what percent will the area be decreased?

(a) 20%

(b) 36%

(c) 40%

(d) 60%

(e) 64%

Practice Set 8

4. $\boxed{a|b|c|d|e}$ is a *mean number strip* if b is the arithmetic mean of a and c, c is the mean of b and d, and d is the mean of c and e. If $\boxed{a|b|c|d|e}$ is a mean number strip, which of the following is an expression for e in terms of a and b?

(a) $\dfrac{a + b + c + d}{8}$

(b) $\dfrac{a + b + c + d}{4}$

(c) $\dfrac{a + b}{2}$

(d) $4b - 3a$

(e) $7b - 10a$

5. In the starlike figure below, what is the value of x?

(a) 20°

(b) 25°

(c) 30°

(d) 40°

(e) 50°

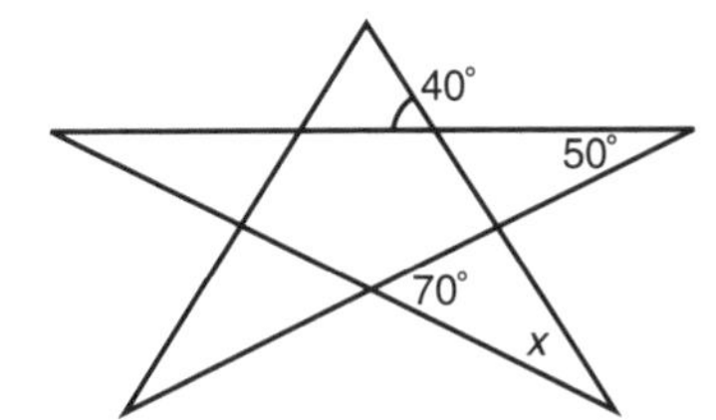

6. If $x^2 - 9 = (x + p)(x - p)$ for all values of x, which of the following could be the value of p?

(a) 1

(b) 2

(c) 3

(d) 4

(e) 9

7. If $2x - 8 = 1$, what is the value of $x - 4$?

(a) $-\frac{1}{2}$

(b) $\frac{1}{2}$

(c) 4

(d) 8

(e) $8\frac{1}{2}$

8. Jim earns x dollars in h hours. How many dollars will he earn in $h + 20$ hours?

(a) $\frac{20x}{h}$

(b) $x + \frac{20x}{h}$

(c) $21x$

(d) $\frac{xh}{h + 20}$

(e) $(h + 20)x$

9. The vertex-edge graph below shows the number of minutes it takes Brianna to bicycle between the different stops on her delivery route. What is the least amount of time, in minutes, it could take her to visit each stop if she must begin and end at vertex A?

(a) 6

(b) 13

(c) 15

(d) 18

(e) 24

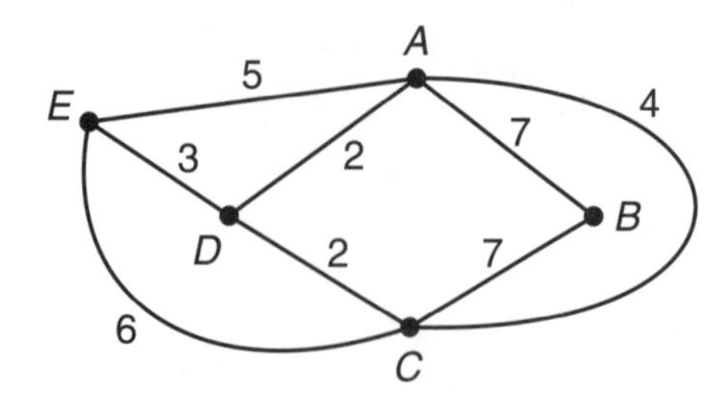

10. If $18 - 3(5 - x) = x - 7$, then $x =$

(a) -8

(b) -5

(c) -1

(d) 5

(e) 10

Practice Set 8

Test Taking Tip

When considering relations among similar shapes, be sure to identify the dimensions of the aspects being compared.

When two shapes are similar with a scale factor of k, corresponding linear measures are related by the factor k. However, when areas or volumes are compared, the areas are related by a factor of k^2, and the volumes are related by a factor of k^3. Be sure you know whether the measures being compared are for line segments, plane regions, or space regions.

Example Look back at Item 1 on page 116. The triangles are given as similar and the scale factor is found to be $\frac{3}{2}$. Now you seek the perimeter—a number that is measured in linear units—of the second triangle. Since the first triangle has a perimeter of 24, the other one has perimeter $\frac{3}{2} \cdot 24 = 36$. So the answer is (d). What multiplier would you use if you wanted the area of the second triangle, given the area of the first?

- Find, if possible, another test item in the practice set for which this strategy might be helpful. Try it.
- Keep this strategy in mind as you work on future practice sets.

Practice Set 9

Solve each problem. Then record the letter that corresponds to the correct answer.

1. The endpoints of a diagonal of a square are $(-2, 2)$ and $(1, 5)$. What is the area of the square?

(a) 3

(b) $3\sqrt{2}$

(c) 9

(d) 12

(e) 18

2. Frances can type 100 pages in h hours. At this rate, how many pages can she type in m minutes?

(a) $\frac{mh}{60}$

(b) $\frac{60m}{h}$

(c) $\frac{100m}{60h}$

(d) $\frac{60h}{100m}$

(e) $\frac{100h}{60m}$

3. Suppose in the figure below, the measure of $\angle 5$ is 120°. Find the measure of $\angle 2$.

(a) 30°

(b) 45°

(c) 60°

(d) 90°

(e) 120°

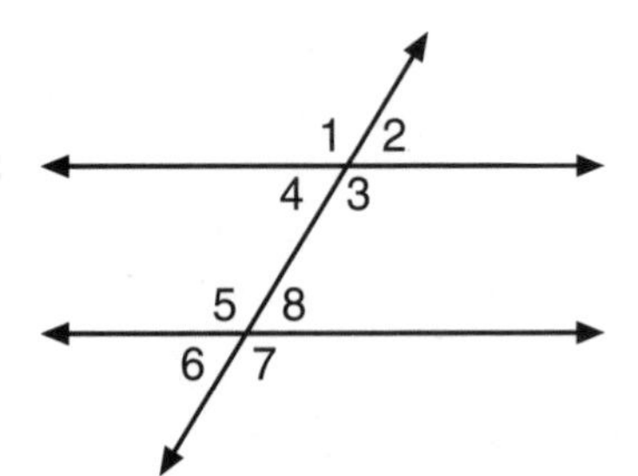

4. Which of the following is a solution to the equation $|3x - 6| = 3$?

(a) -2

(b) -1

(c) 0

(d) 1

(e) 2

5. At a speed of 48 miles per hour, how many minutes will it take to drive 40 miles?

(a) $\frac{5}{8}$

(b) $\frac{8}{5}$

(c) 32

(d) 50

(e) 1,920

6. If $a < 0$ and $b < 0$, which of the following must be less than 0?

I. $a + b$ II. $a \cdot b$ III. $-\frac{a}{b}$

(a) None

(b) I and II only

(c) I and III only

(d) II and III only

(e) I, II, and III

7. If $\frac{n}{6} = 8$, what is the value of $\frac{n}{8}$?

(a) 6

(b) $\frac{4}{3}$

(c) $\frac{3}{4}$

(d) $\frac{1}{6}$

(e) 2

8. For nonzero numbers x, y, and z, $3x = 4y$ and $12y = 5z$. What is $\frac{z}{x}$?

(a) 4

(b) $\dfrac{16}{5}$

(c) $\dfrac{9}{5}$

(d) $\dfrac{6}{5}$

(e) $\dfrac{5}{16}$

9. On standardized tests like the ACT and SAT, every question is worth the same amount: one point. Assume there are five choices, A–E, for each item. Suppose you got bogged down on some questions and, with a minute left, you still have ten questions to answer. If you guess on each question and there is no guessing penalty, what is your expected score on the last ten items of the test?

(a) 1

(b) 2

(c) 3

(d) 4

(e) 5

10. The following sketch shows a shed wall. What is the area of the surface of this wall, in square feet?

(a) 24

(b) 56

(c) 64

(d) 72

(e) 504

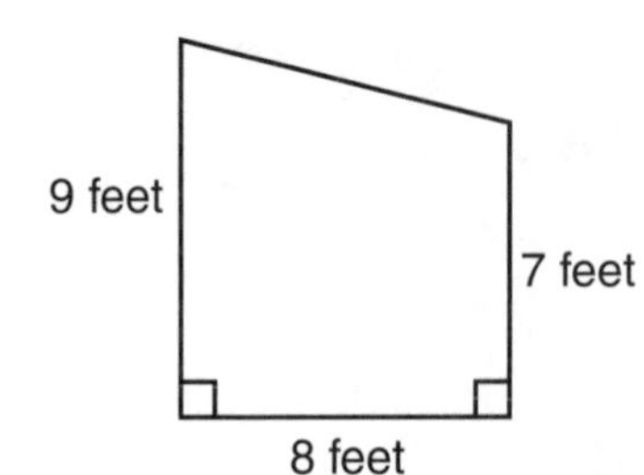

Practice Set 9

Test Taking Tip

Pay attention to units used in a problem.

Often the answer to a question must be in units different than those used in the statement of the problem. Be sure to read problems carefully and make the necessary conversions.

Example Look back at Item 5 on page 121. The speed is given in miles per hour, but the question asks for an answer in minutes. First determine the time it will take in hours, then do the necessary conversion. Since $t = \frac{d}{r}$, $d = 40$ miles, and $r = 48$ mph, you have $t = \frac{40}{48}$ hours. But, $\frac{40}{48}$ hours $= \frac{40}{48}$ hours $\cdot \frac{60 \text{ min}}{1 \text{ hour}} = \frac{2{,}400}{48}$ min or 50 minutes. So the correct answer is (d).

- Find, if possible, another test item in the practice set for which this strategy might be helpful. Try it.
- Keep this strategy in mind as you work on future practice sets.

Practice Set 10

Solve each problem. Then record the letter that corresponds to the correct answer.

1. The side length of a square is w units. Its area is 8 square units. Find the area, in square units, of a square if its side is $4w$ units long.

(a) 16

(b) 25

(c) 32

(d) 128

(e) 168

2. In the figure below, O is the center of square $STEP$ with sides parallel to the axes. If the sum of the coordinates of T is 12, what is the sum of the coordinates of E?

(a) -12

(b) -6

(c) 0

(d) 6

(e) 12

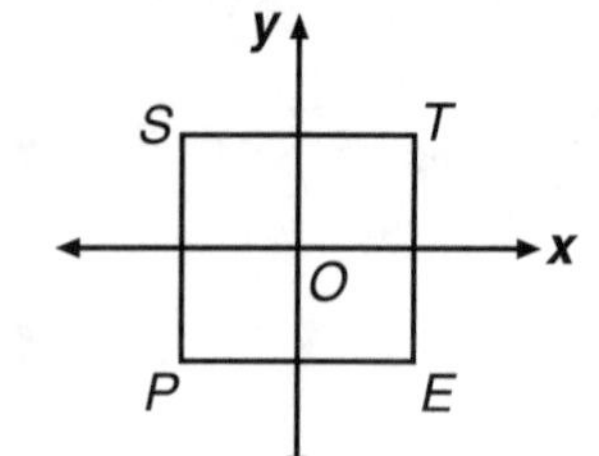

3. If a car can be driven 45 miles on 2 gallons of gasoline, how many gallons of gasoline will it take to travel 315 miles?

(a) 3.5

(b) 7

(c) 9

(d) 14

(e) 16

4. If $y = x + 4$, what is the value of $|y - x| + |x - y|$?

(a) 0

(b) 2

(c) 4

(d) 8

(e) 16

5. Which is the equation of a line parallel to $3x - 4y = 12$?

(a) $y = 3x + 5$

(b) $y = -\frac{3}{4}x + 8$

(c) $y = \frac{3}{4}x - 6$

(d) $y = \frac{4}{3}x - 3$

(e) $y = -\frac{4}{3}x + 12$

6. For the rectangular concrete block shown, measures l, w, and h are unequal. The diagonals of the faces of the block have how many different lengths?

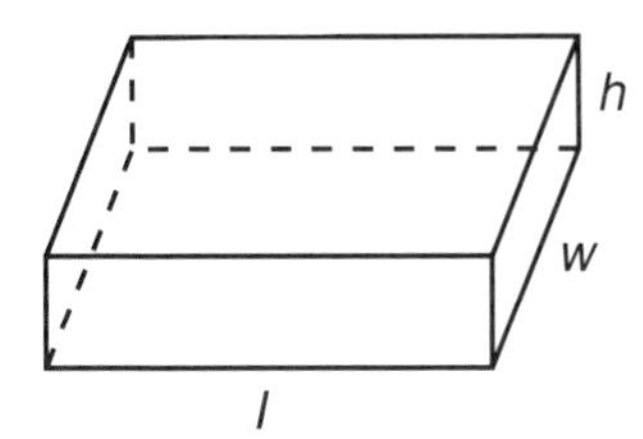

(a) One

(b) Two

(c) Three

(d) Four

(e) Twelve

7. Suppose on a standardized multiple-choice test with five choices, each question is worth the same amount: 1 point. To counter guessing, a guessing penalty of $\frac{1}{4}$ of a point is applied to each incorrect answer. Suppose on each of the last ten items, you were able to eliminate two choices, but did not have time to complete solutions of any of the items. If you chose to answer none of them, your score on the ten items is 0. If you guess on each of the questions, what is your expected score on the ten items?

   (a) 0
   (b) $\frac{1}{3}$
   (c) $\frac{2}{3}$
   (d) $\frac{5}{3}$
   (e) $\frac{15}{4}$

8. If n is a negative integer, how do $(-2)^{2n}$ and $(-2)^{2n+1}$ compare?

   (a) $(-2)^{2n} > (-2)^{2n+1}$
   (b) $(-2)^{2n} < (-2)^{2n+1}$
   (c) $(-2)^{2n+1} \geq (-2)^{2n}$
   (d) $(-2)^{2n} = (-2)^{2n+1}$
   (e) Cannot be determined from the given information.

9. Suppose that $\triangle BCD$ is isosceles, $CD = 8\sqrt{2}$, and $m\angle A = 30^\circ$. What is the length of $\overline{AB}$?

   (a) $4\sqrt{6}$
   (b) 8
   (c) $8\sqrt{2}$
   (d) $8\sqrt{3}$
   (e) 16

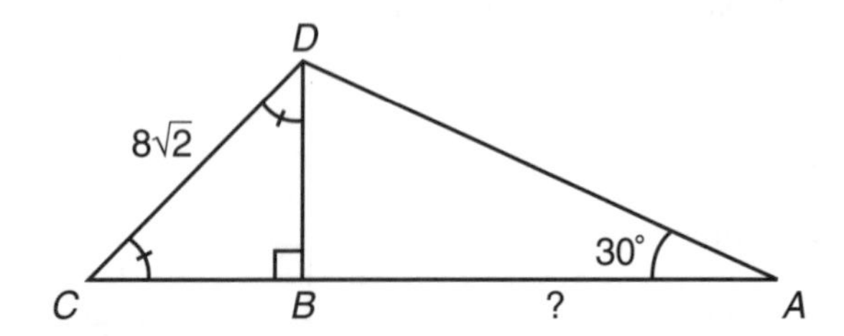

10. A bag contains 10 marbles, 3 of which are red, 2 of which are blue, and 5 of which are yellow. You reach in and draw two marbles. What is the probability that both marbles are blue?

   (a) $\frac{2}{100}$
   (b) $\frac{2}{90}$
   (c) $\frac{4}{100}$
   (d) $\frac{28}{90}$
   (e) $\frac{4}{20}$

Practice Set 10

Test Taking Tip

If all else fails, it pays to make an educated guess.

Previous Test Taking Tips have offered strategies for working efficiently to produce correct answers to multiple-choice questions. Because of time constraints or the nature of some questions on standardized tests, sometimes you may not be able to produce an answer in the allotted time. In these cases, it is to your benefit to make an educated guess, after first eliminating one or two obviously incorrect choices. The ACT has no penalty for guessing. On the SAT, the penalty for guessing is a $\frac{1}{4}$-point or a $\frac{1}{3}$-point reduction for each incorrect answer, depending on the number of choices.

Example Suppose that a tenth-grader answered correctly 10 of the 20 questions from the last two practice sets. If he did not answer any other questions, his total score across the two sets would be 10.

(a) If he guessed at the answer to each of the remaining questions, on average he would get $10\left(\frac{1}{5}\right) = 2$ correct. His score then across the two sets would be 12.

(b) If, before guessing, he was able to eliminate two choices for each item, then, on average, his score would be $10 + 10\left(\frac{1}{3}\right) = 13\frac{1}{3}$ or, when rounded, 13 points for the two sets.

(c) If in Part b a guessing penalty is applied, his score on average would be $10 + 10\left(\frac{1}{3}\right) - \frac{1}{4}\left(10 \cdot \frac{2}{3}\right) = 11\frac{2}{3}$ or, when rounded, 12 points.

- For what items on this practice set might this strategy have been useful for you?
- Keep this strategy in mind as a last resort as you work on future problems of this type.

Solutions

Solutions to Check Your Understanding

Check Your Understanding 1.1, p. 8

1. **a.** $h = 5{,}000 - 250t$ where h is the height, in meters, and t is the time, in minutes; $NEXT = NOW - 250$ (Start at 5,000).

   b. $n = 4 \cdot 2^x$ where n is the number of families notified at that stage and x is the stage number; $NEXT = 2NOW$ (Start at 4).

2. **a.**

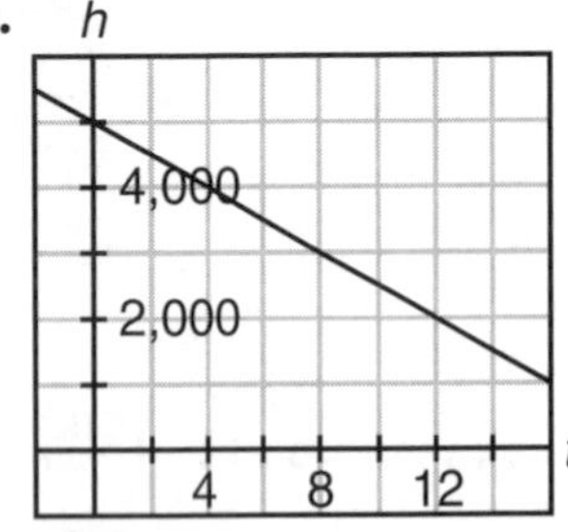

b.

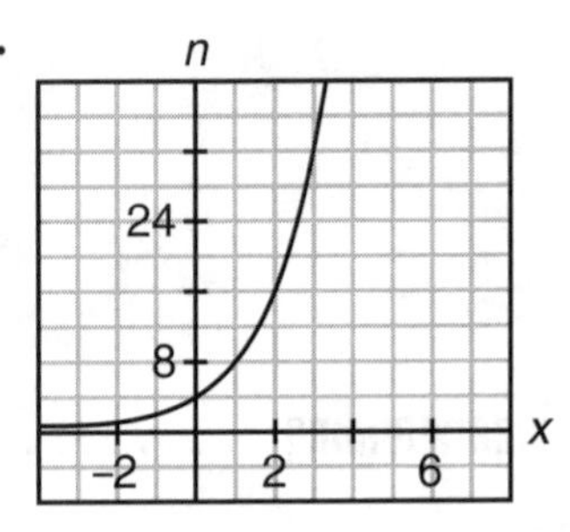

3. **a.** $1{,}000 = 5{,}000 - 250t$. After 16 minutes the plane reaches 1,000 meters.

   b. $512 = 4 \cdot 2^x$; 512 families will be notified at stage 7.

4. **a.** I: $NEXT = 3NOW$ (Start at 2)

   II: $NEXT = NOW + 4$ (Start at 2)

   b. I: $y = 2(3^x)$

   II: $y = 4x + 2$

Check Your Understanding 1.2, p. 12

1. **a.** $7.3r + 5.7s = 300$; $r + s = 45$; r is the number of minutes on a rowing machine and s is the number of minutes on a stair machine.

   b. 27.2 minutes on the rowing machine and 17.8 minutes on the stair machine

2. The display costs are the same after 1.25 hours. The cost is $10.00.

3. **a.** $(x, y) = (5, 1)$ **b.** $(x, y) = (5.2, 1.8)$ **c.** $(x, y) = (1, 3)$

   d. $(x, y) = (2, -1)$ **e.** $(x, y) = (1, 0)$ **f.** $(x, y) = (-1, -1)$

Check Your Understanding 1.3, p. 14

1. **a.** $s = kh^2$ where s is the strength of the bone and h is the height of the animal.

   b. $w = kh^3$ where w is the weight of the animal and h is the height of the animal.

c. $l = \frac{k}{d^2}$ where l is the loudness of the sound and d is the distance from the speaker.

2. $V = \frac{64{,}000}{l}$ where l is the length in centimeters and V is the number of vibrations per second. A wire needs to be 266.67 centimeters long to vibrate 240 times per second.

3. $f = \frac{k}{\sqrt{w}}$ is an equation for the inverse variation; f is 254 when w is 4.

Check Your Understanding 1.4, p. 17

1. a. $x = 2$ or $x = -6$ **b.** $x = 1.25$ or $x = -1$

c. $x = 2$ **d.** No solution

2. The lengths of the sides are x and $x + 5$, so $A = x(x + 5)$. The length is 20 meters, and the width is 15 meters.

3. a. $x = \pm 4$ **b.** $x = \pm 3\sqrt{3}$ **c.** $x = \pm 6$

d. No solution **e.** $x = \pm 2\sqrt{2}$ **f.** $x = \pm\sqrt{\frac{9}{2}} = \pm\frac{3}{\sqrt{2}}$

4. a. Opens upward, the axis of symmetry is $x = \frac{5}{4}$, the y-intercept is -12, and the x-intercepts are $-\frac{3}{2}$ and 4.

b. Opens downward, the axis of symmetry is $x = 0$, the y-intercept is 12, and the x-intercepts are $\sqrt{6}$ and $-\sqrt{6}$.

c. Opens upward, the axis of symmetry is $x = \frac{1}{6}$, the y-intercept is -2, and the x-intercepts are 1 and $-\frac{2}{3}$.

d. Opens downward, the axis of symmetry is $x = 1$, the y-intercept is 1, and the x-intercepts are approximately -0.4142 and 2.4142.

Check Your Understanding 1.5, p. 19

1. a. x^6 **b.** a^{12} **c.** x^{-6} or $\frac{1}{x^6}$ **d.** c^6

e. 3^3 or 27 **f.** 3^{-3} or $\frac{1}{27}$ **g.** x^{10} **h.** a^2b^{-3} or $\frac{a^2}{b^3}$

2. a. $2\sqrt{6}$ **b.** $9\sqrt{3}$ **c.** $3\sqrt[3]{9}$ **d.** $2\sqrt[3]{4}$

e. $2ab^2\sqrt{2ab}$ **f.** $4xy^2\sqrt[3]{x^2}$ **g.** $6\sqrt{2}$ **h.** $2\sqrt[3]{9}$

3. a. $\frac{1}{t^8}$ **b.** 16 **c.** $-\frac{1}{8}$ **d.** $\frac{1}{16}$

e. a^{10} **f.** c^{13} **g.** x^0 or 1 **h.** b^{-3} or $\frac{1}{b^3}$

Check Your Understanding 2.1, p. 21

1. **a.**

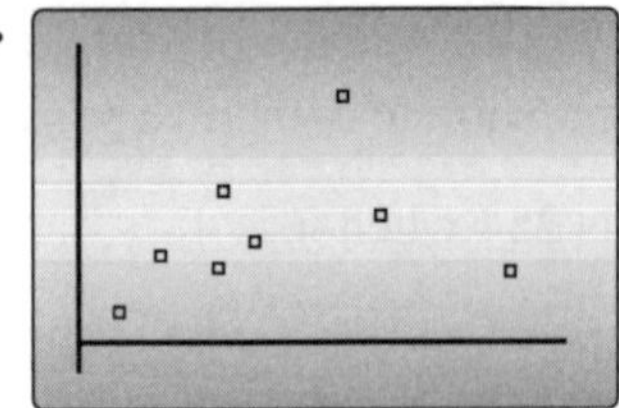

The linear association is positive and moderately strong.

b. 0.274 approximately

c. The point should increase the correlation coefficient since it stretches the scatterplot up and to the right. The new correlation coefficient is approximately 0.655.

d. The centroid is (388.33, 752.78) (with the additional point included). When 388.33 is substituted into the regression equation $y = 1.39x + 214.11$ the result is 753.9, close enough when round off is considered. Thus, the centroid in on the regression line.

2. **a.** This is an influential point since it is somewhat out of the pattern. Deleting it increases the linear association.

b. The correlation coefficient increases to 0.865.

c. You can determine that they are influential points by observing that they are outliers on a scatterplot of the data. The point (550, 480) decreases the correlation coefficient while the point (710, 1400) increases the correlation coefficient.

3. **a.** $y = 0.614x + 458.12$. An increase of 1 calorie in the food corresponds to an increase of 0.614 mg sodium in that food.

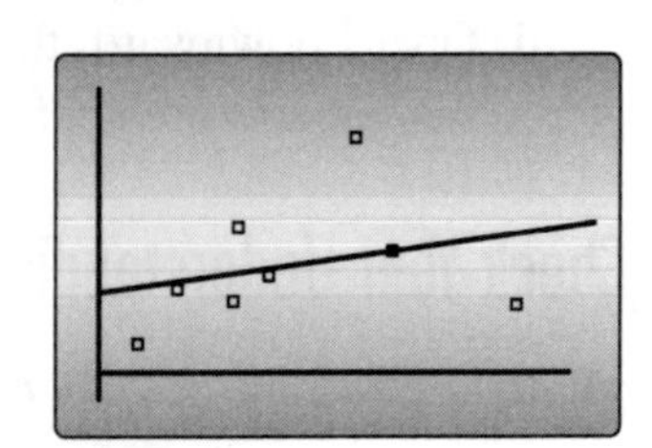

b. The residual is $480 - (0.614 \cdot 550 + 458.12) = -315.82$.

Check Your Understanding 2.2, p. 25

1. **a.** $P = \frac{3}{17} \approx 0.176$ **b.** $P = \frac{1}{17} \approx 0.059$ **c.** $P = \frac{1}{16} \approx 0.0625$

d. $P = \frac{1}{2{,}652} \approx 0.00038$ **e.** $P = \frac{1}{32} \approx 0.03125$

2. The expected value of the game is 18.75 cents.

3. a. The distribution is as follows for 1 – 10 boxes:

Number of Boxes	1	2	3	4	5	6	7	8	9	10
Probability of Getting First Prize	0.2	0.16	0.128	0.1024	0.0819	0.0655	0.0524	0.04194	0.0336	0.0268

b. The expected value of the distribution is $\frac{1}{\frac{1}{5}}$ or 5 boxes.

c. It is not a rare event since $P(< 9) \approx 0.83$, so $P(\geq 9) \approx 0.17$ which is greater than 0.05.

Check Your Understanding 3.1, pp. 27–28

1. a. $y = -\frac{1}{3}x + \frac{11}{13}$ **b.** $y = \frac{4}{9}x - \frac{11}{9}$

2. a. $PQ = \sqrt{13}$, $QR = \sqrt{13}$, $PR = \sqrt{26}$

- $\triangle PQR$ is isosceles since $PQ = QR$.
- $\triangle PQR$ is a right triangle since $(\sqrt{13})^2 + (\sqrt{13})^2 = (\sqrt{26})^2$. Thus the Pythagorean Theorem is satisfied.

b. Midpoint of $\overline{PQ}$: $(-3.5, 1)$; midpoint of $\overline{QR}$: $(-4, 3.5)$; midpoint of $\overline{PR}$: $(-2.5, 2.5)$

3. a. $y = 0.25x + 5$ **b.** $y = -4x + 3$ **c.** $y = -\frac{1}{3}x - \frac{5}{3}$ **d.** $y = -\frac{9}{5}x + 11$

Check Your Understanding 3.2, p. 30

1. a. $\begin{bmatrix} 1 & 0 \\ 0 & -1 \end{bmatrix}$ **b.** $\begin{bmatrix} -1 & 0 \\ 0 & -1 \end{bmatrix}$ **c.** $\begin{bmatrix} 0 & -1 \\ -1 & 0 \end{bmatrix}$

d. $\begin{bmatrix} -1 & 0 \\ 0 & 1 \end{bmatrix}$ **e.** $\begin{bmatrix} 0 & 1 \\ -1 & 0 \end{bmatrix}$ **f.** $\begin{bmatrix} 0.4 & 0 \\ 0 & 0.4 \end{bmatrix}$

2. The images of $A(2, 5)$ and $B(8, -3)$ are $A'(9, 22.5)$ and $B'(36, -13.5)$. The slope of the line through A and B is $-\frac{8}{6} = -\frac{4}{3}$. Similarly, the slope of the line through A' and B' is $\frac{-13.5 - 22.5}{36 - 9} = \frac{-36}{27} = \frac{-4}{3}$ and the slopes of the two lines are identical. Thus, the lines are parallel.

Check Your Understanding 3.3, pp. 34–35

1. Approximately 173.2 feet

2. a. $q = \sqrt{p^2 + r^2}$ **b.** $\sin R = \frac{\sqrt{55}}{8}$, $\cos R = \frac{3}{8}$, and $\tan R = \frac{\sqrt{55}}{3}$

c. The missing side has a length of 24. Thus, 25, 24, and 7, as well as any positive multiple of these three numbers, could be the lengths of the sides. Two examples are 50, 48, and 14 and 100, 96, and 28.

3.

Measure of angle	30°	45°	75°
Sine of angle	0.5	$\frac{1}{\sqrt{2}} = \frac{\sqrt{2}}{2}$	0.9659
Cosine of angle	$\frac{\sqrt{3}}{2}$	$\frac{1}{\sqrt{2}} = \frac{\sqrt{2}}{2}$	0.2588
Tangent of angle	$\frac{1}{\sqrt{3}} = \frac{\sqrt{3}}{3}$	1	3.7321

4. a. $\frac{\pi}{6}$ **b.** $\frac{\pi}{4}$ **c.** $\frac{\pi}{3}$

d. $\frac{\pi}{2}$ **e.** π **f.** $\frac{3\pi}{2}$

5. a. 30° **b.** 60° **c.** 45°

d. 135° **e.** 120° **f.** 330°

6. a. $\frac{8\pi}{3} \frac{\text{rad}}{\text{sec}}$ **b.** $64\pi \frac{\text{in}}{\text{sec}}$

7. a. The graphs of $y = 3\sin x$ and $y = \sin x$ have the same period and intersect the x-axis in the same points, but each value of y for $y = 3\sin x$ is three times the corresponding value of y for $y = \sin x$.

b. Each point on the graph of $y = -3\cos x$ is 3 times as far from the x-axis and on the opposite side of the x-axis as the corresponding point on $y = \cos x$.

Check Your Understanding 4.1, pp. 39–40

1. a. $\begin{bmatrix} 1 & 1 & 3 \\ 6 & 7 & 3 \end{bmatrix}$ **b.** $\begin{bmatrix} -1 & -11 & -1 \\ 0 & -3 & 5 \end{bmatrix}$ **c.** $\begin{bmatrix} -2 & -12 & -4 \\ -6 & -10 & 2 \end{bmatrix}$ **d.** $\begin{bmatrix} -10 & -8 & 8 \\ 4 & 11 & 1 \end{bmatrix}$

2. a.

$$\begin{array}{c} \\ 28'' \\ 30'' \\ 32'' \\ 34'' \end{array}\begin{array}{c} \begin{array}{cccccc} L & TH & P & W & J & CK \end{array} \\ \begin{bmatrix} 6 & 0 & 1 & 6 & 3 & 7 \\ 14 & 4 & 2 & 2 & 0 & 4 \\ 9 & 0 & 3 & 2 & 0 & 1 \\ 6 & 5 & 8 & 2 & 4 & 5 \end{bmatrix} \end{array}$$

- 15 pairs
- JNCO

b.

$$\begin{array}{c} \\ 28'' \\ 30'' \\ 32'' \\ 34'' \end{array}\begin{array}{c} \begin{array}{cccccc} L & TH & P & W & J & CK \end{array} \\ \begin{bmatrix} 6 & 0 & 1 & 6 & 3 & 7 \\ 14 & 12 & 2 & 2 & 0 & 12 \\ 9 & 12 & 3 & 2 & 0 & 12 \\ 6 & 5 & 8 & 2 & 4 & 5 \end{bmatrix} \end{array}$$

c. ■ The product of the sales matrix and the profit matrix is $\begin{bmatrix} 210 \\ 451 \\ 491 \\ 225.50 \end{bmatrix}$

The store made \$225.50 profit on the largest size sold and \$210 on the smallest size sold.

- \$1,377.50

3. a.

$$\begin{array}{c} \\ A \\ B \\ C \\ D \end{array}\begin{array}{c} \begin{array}{cccc} A & B & C & D \end{array} \\ \begin{bmatrix} 3 & 2 & 2 & 2 \\ 2 & 3 & 2 & 2 \\ 2 & 2 & 3 & 2 \\ 2 & 2 & 2 & 3 \end{bmatrix} \end{array}$$

Paths from *D* to *C*: *D–B–C*, *D–A–C*.

b. 24 three-stage paths start and finish at the same point; 84 three-stage paths start and finish at different points.

Check Your Understanding 4.2, p. 43

1.

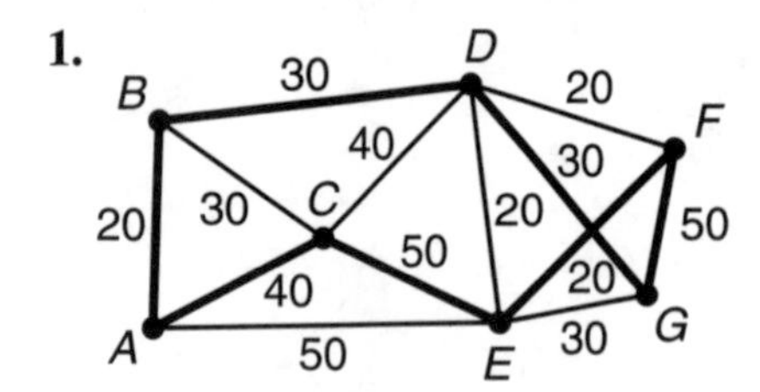

2.

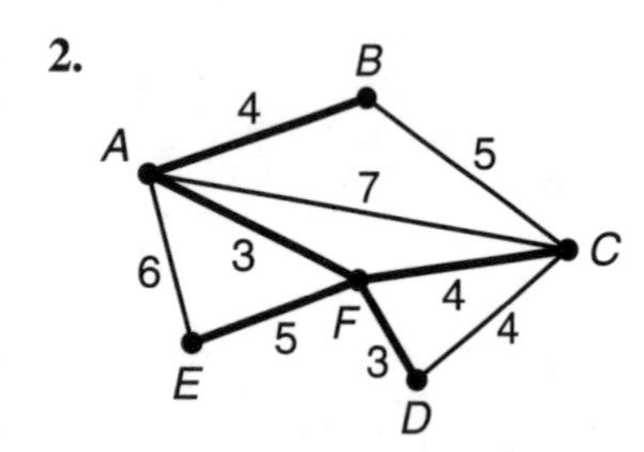

3.

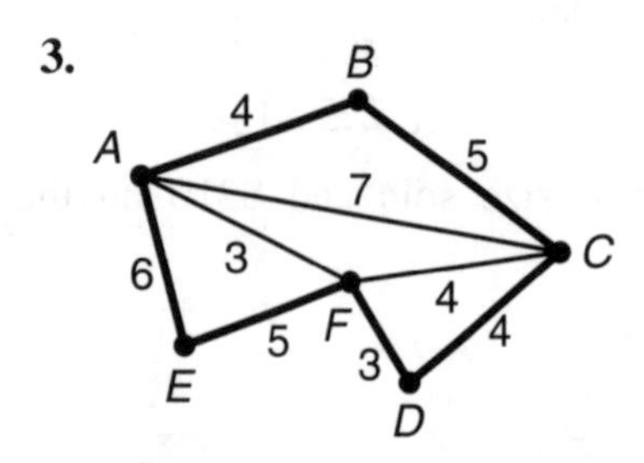

For this Hamiltonian circuit, the length is 27 miles.

4. Yes, one path covering all roads is *A–B–C–F–A–C–D–F–E–A*

Solutions to Exercise Sets

Exercise Set 1, pp. 46–47

1. a. $y = 20 + 0.075x$

b. $NEXT = NOW + 0.075$ (Start at 20)

c. 242 minutes

d. $47.90

e. A maximum of 466 minutes

f. Less than 400 minutes

2. a. $(x, y) = (2, 3)$

b. $(x, y) = (-1, 3)$

c. $(x, n) = (2, \frac{5}{3})$

3. a. $AB = 5$, $BC = \sqrt{37} \approx 6.08$, $CA = \sqrt{18} \approx 4.24$

b. $\triangle ABC$ is not a right triangle since the Pythagorean Theorem is not satisfied.

c. Midpoint of $\overline{AB}$: (2.5, 0); midpoint of $\overline{BC}$: (1, −1.5); midpoint of $\overline{CA}$: (−0.5, 0.5)

d. $\overleftrightarrow{AB}: y = -\frac{4}{3}x + \frac{10}{3}$; $\overline{BC}: y = -\frac{1}{6}x - \frac{4}{3}$; $\overleftrightarrow{CA}: y = x + 1$

4. a. $x \leq 2$

b. $x \geq 2$ or $x \leq -2$

c. $x \geq -7$

d. $x \geq 0$

e. $-\sqrt{6} < x < \sqrt{6}$

5. a. 5

b. $\frac{1}{q^2}$

c. $\frac{1}{2^9}$

d. $\frac{1}{4^2} = \frac{1}{16}$

e. x^4y^8

f. $\frac{1}{x^6}$

6. a. 219.8 feet

b. 69.28 feet

c. 80 feet

7. a. Let N denote the number of seats and r the number of rows, then $N = r(r - 16)$.

b. 43 rows

c. 55 rows and 39 seats in each row

8. a. The x-intercepts are 1 and 3. The y-intercept is 3.

b. $x = 2$

c. The graph opens upward since the coefficient of x^2 is positive.

d.

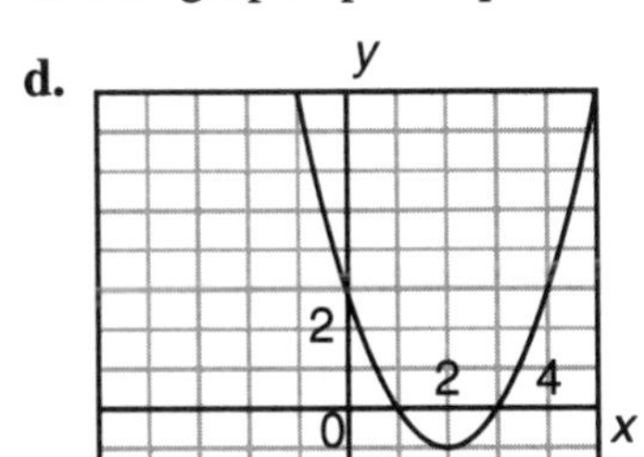

9. **a.** $\begin{bmatrix} -1 & 0 \\ 0 & 1 \end{bmatrix}$ **b.** $\begin{bmatrix} 2 & 0 \\ 0 & 2 \end{bmatrix}$ **c.** $\begin{bmatrix} 0 & 1 \\ 1 & 0 \end{bmatrix}$ **d.** $\begin{bmatrix} 0 & -1 \\ 1 & 0 \end{bmatrix}$

10. **a.** $\left(\frac{31}{36}\right)^3 \left(\frac{5}{36}\right) \approx 0.0887$ **b.** $\frac{1}{\frac{5}{36}} = \frac{36}{5} = 7.2$ rolls

Exercise Set 2, pp 48–49

1. **a.** $p = 0.01(2^{n-1})$ **b.** $NEXT = 2NOW$ (Start at 0.01) **c.** $x = 11$ days

   d. \$5,242.88 **e.** 11 days

2. **a.** Metro Cab: $y = 2.50 + 0.20x$
   National Cab: $y = 1.50 + 0.30x$

   b. National Cab **c.** Metro Cab **d.** 10 miles

3. **a.** $A'(2, 4)$, $B'(-6, -5)$, $C'(1, -4)$

   b. $A'(-2, -4)$, $B'(6, 5)$, $C'(-1, 4)$

   c. $A'(-4, 2)$, $B'(5, -6)$, $C'(4, 1)$

4. **a.** $x = \pm\sqrt{48} = \pm 4\sqrt{3}$ **b.** $x = \pm\sqrt{3}$

   c. $x = \pm\sqrt{11}$ **d.** $x = \pm 3$

5. **a.** $\begin{bmatrix} 2 & 16 \\ -10 & 3 \end{bmatrix}$ **b.** $\begin{bmatrix} 10 & 2 \\ 14 & -3 \end{bmatrix}$ **c.** $\begin{bmatrix} -132 & 69 \\ -8 & 14 \end{bmatrix}$

6. **a.** The coordinates of point C are $(5, 5)$. $AB = 4$. $BC = 3$.

   b. $\overleftrightarrow{CA} : y = \frac{3}{4}x + \frac{5}{4}$; $\overleftrightarrow{AB} : y = 2$; $\overleftrightarrow{BC} : x = 5$

7. **a.** 310.42 feet **b.** 254.3 feet

8. **a.** Responses may vary. One minimal spanning tree is shown below. **b.** 12

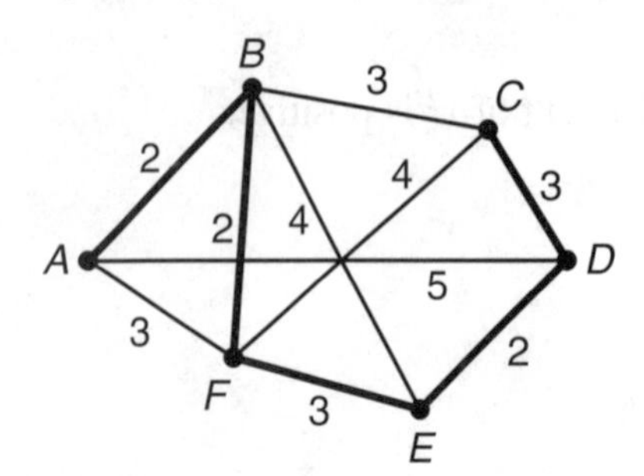

c. 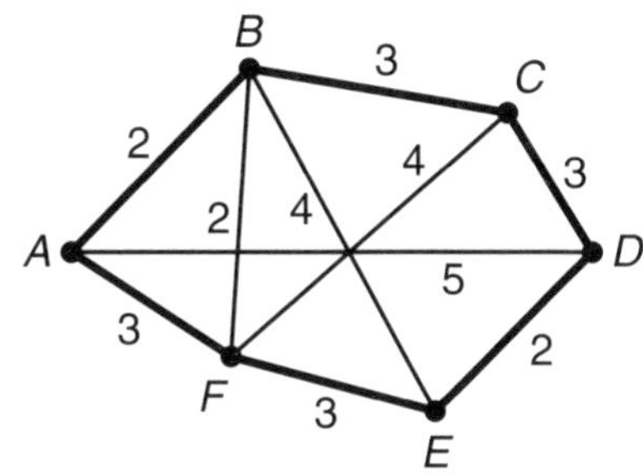

d. 16

9. a. Let A denote the area and w the width, then $A = w(w + 6)$.

b. $w \approx 7.95$ feet

c. 60 feet

10. a. Yes. The data is reasonably modeled by a line since the data points seem to cluster roughly in a linear fashion.

b. $y = 1.87x - 2.63$

c. The slope of the regression line is 1.87. This means that for every one year increase in age, the children tended to watch the video about 1.87 minutes longer.

d. The residual is -0.11.

Exercise Set 3, pp. 50–51

1. a. $6x - y = 3$

b. $2x + 2y = -5$

c. $6x + 12y = 1$

d. $6x - y = 15$

e. $x - 15y = 10$

f. $5x + 3y = 30$

2. a. $(x, y) = (1, 6)$

b. $(x, y) = (2, 1)$

c. $(x, y) = \left(2, \frac{4}{3}\right)$

3. a. $y = x^2$

b. $y = x^2 + 2$

c. $-x^2 + 2$

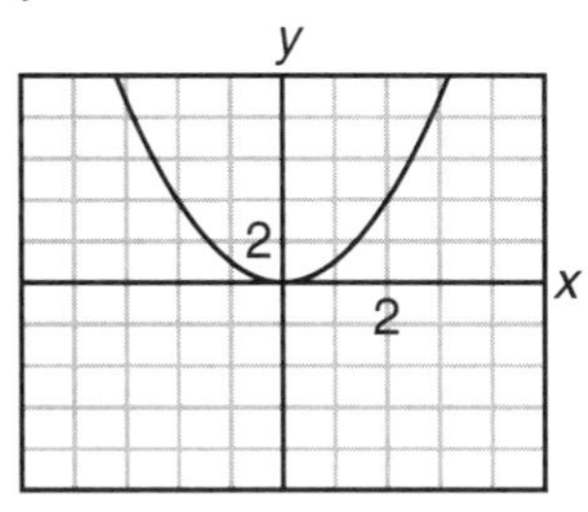

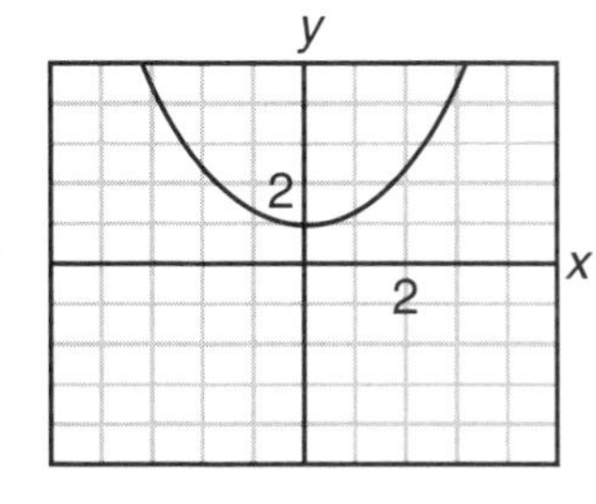

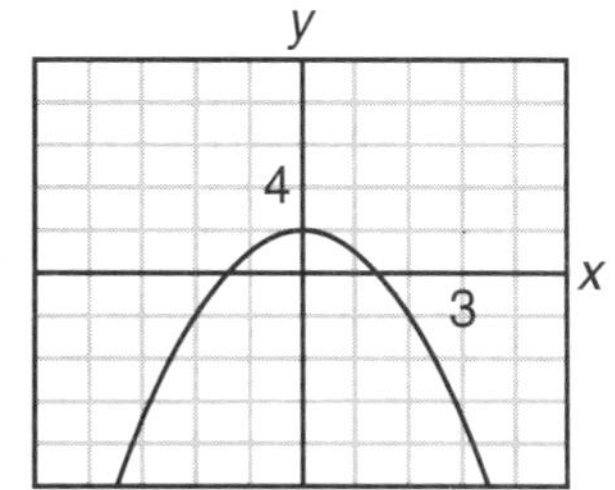

4. 200 adult's and 300 children's tickets

Sample:

5. a.

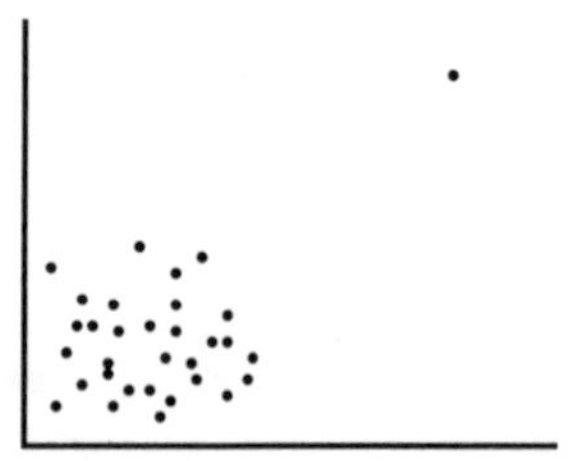

Sample:

b.

6. a. $x = -8$ or $x = 3$ **b.** $x = 2$ or $x = -\frac{1}{3}$ **c.** $x = 2$

7. a–b.

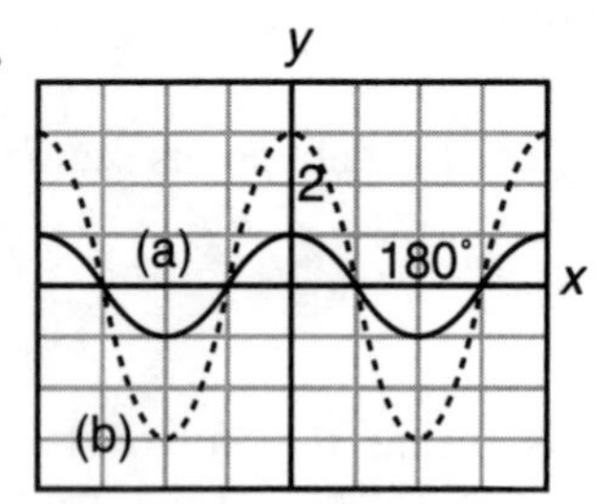

c–d.

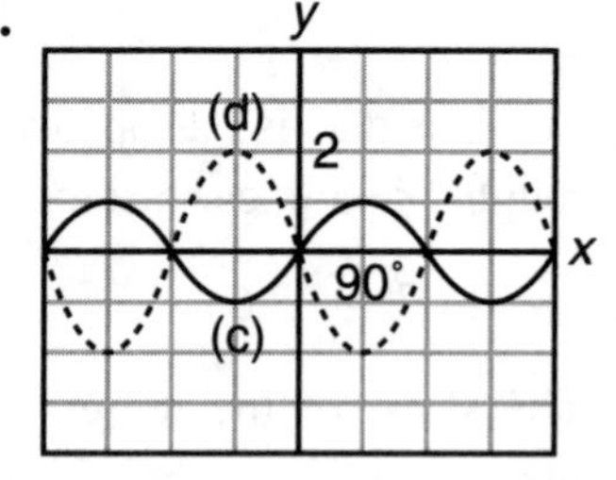

8. a.

x	1	2	3	4	n
$P(x)$	0.6	0.6(0.4)	$0.6(0.4)^2$	$0.6(0.4)^3$	$0.6(0.4)^{n-1}$

b.

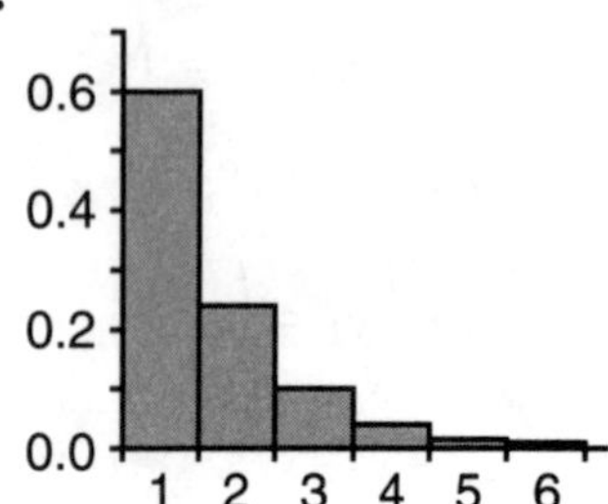

c. About 0.015

d. No. $P(x < 4) = 0.6 + 0.6(0.4) + 0.6(0.4)^2 \approx 0.936$. So, $P(x \geq 4) = 1 - 0.936 = 0.064$, which is greater than 5%.

9. a.

Radius (cm)	1	2	3	4	7	r
Volume (cm³)	10π	40π	90π	160π	490π	$10r^2\pi$

b.

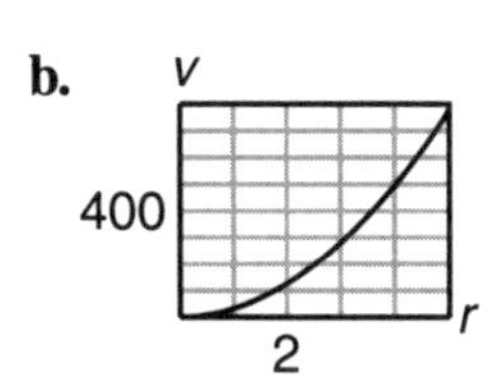

c. As the radius increases, the volume increases at an increasing rate.

d. The volume is multiplied by 9.

10. a. 17 feet **b.** 2 feet **c.** 17 feet

Exercise Set 4, pp. 52–53

1. a. $y = x - 3$ **b.** $y = \frac{3x + 3}{2}$ **c.** $y = \frac{2x - 1}{3}$

d. $y = \frac{8x}{3}$ **e.** $y = 3x$ **f.** $y = \frac{-x - 16}{5}$

2. $x \approx 0.17$, $y \approx 0.32$

3. a. 36 **b.** $4\sqrt{6}$ **c.** $\frac{5}{8}$ **d.** $39\sqrt{22}$

e. $36\sqrt{77}$ **f.** 8 **g.** $2x^3y$ **h.** $12r^4\sqrt{2r}$

4. a. The y-intercept is 2 and the x-intercepts are 90° and 270°.

b. The maximum value of 2 occurs at the points (0, 2) and (360, 2).

c. The minimum value of −2 occurs at the point (180, −2).

5. a. $AB = \sqrt{5}$, $BC = 1$, $CA = \sqrt{10}$

b. Midpoint of $\overline{AB}$: (2, 4.5); midpoint of $\overline{BC}$: (3.5, 5); midpoint of $\overline{CA}$: (2.5, 4.5)

c. Slope of $\overline{AB}$: $\frac{1}{2}$; slope of $\overline{BC}$: 0; slope of $\overline{CA}$: $\frac{1}{3}$

d. $x = 3$

6. a. Let P denote the total number of players and r the number of rows, then $P = r(r - 7)$.

b. 16 rows **c.** 8 players **d.** 7 players per row and 14 rows

7. a. Approximately 3.75 feet

b. Approximately 9.27 feet

c. The ladder makes approximately a 66.4° angle with the ground and a 23.6° angle with the building.

8. a. Yes. The data points seem to cluster roughly in a linear fashion.

b. $y = 0.65x + 1.42$

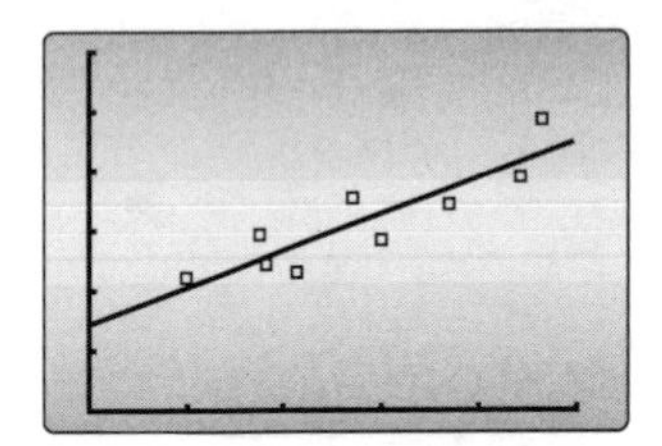

c. From examining the scatterplot and the regression line, you can see that the largest residual is for the point (4.5, 5.1). Its value is $5.1 - (0.65(4.5) + 1.42) = 5.1 - 4.345 = 0.755$.

9. a. $A'(-3, -1), B'(-7, 2), C'(1, 2)$

b. $A'(-1, -2), B'(-5, -5), C'(3, -5)$

c. $A'(-7.5, 2.5), B'(-17.5, -5), C'(2.5, -5)$

10. a. Responses will vary. Possible paths include: *A–B–C, A–B–F–C, A–B–F–D–C, A–E–B–C, A–E–D–C,* and *A–E–F–C*. There are others.

b. There are two paths of the length 14: *A–E–F–C* and *A–E–D–C*.

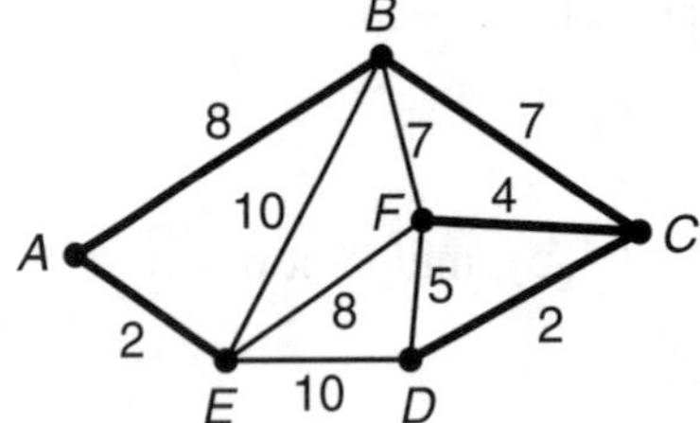

d. Using the minimal spanning tree from Part c, *A–B–C* is the shortest path from *A* to *C*.

Exercise Set 5, pp. 54–55

1. a. 25

b. -5

c. 7

d. $-\frac{4}{5}$

e. $\frac{1}{5}$

f. -25

2. a. $(x, y) = (3, 2)$ **b.** $(x, y) = (-59, 36)$ **c.** $(x, y) = (-3, -2)$

3. 1,200 tickets at $15 and 800 tickets at $25

4. a. $x = \pm\sqrt{3}$ **b.** $x = \pm\sqrt{11}$ **c.** $x = \sqrt[3]{6}$

d. $x = \pm\sqrt{\frac{58}{5}}$ **e.** $x = 4$ **f.** $x = \pm\sqrt{\frac{5}{2}}$

5. a. $\frac{y^2}{x^3}$ **b.** $\frac{1}{t^5}$ **c.** $\frac{1}{a^4}$ **d.** x^6

e. $216x^9$ **f.** $\frac{-2}{x^2}$ **g.** $12x^5$ **h.** $\frac{x^5y^5}{6}$

6. a. $2\sqrt{13} + \sqrt{26} \approx 12.31$ units **b.** 6.5 square units

c. ■ $3(2\sqrt{13} + \sqrt{26}) \approx 36.93$ units **d.** ■ $2\sqrt{13} + \sqrt{26} \approx 12.31$ units

■ $9(6.5) = 58.5$ square units ■ 6.5 square units

7. a. Yes, the data is reasonably modeled by a line since the points cluster about a line. $y = 1.67x + 0.156$

b. 0.91

c. The most influential point is (5.7, 7.5). If the point is removed, the value of r should increase since this point is far away from the regression line. In fact, r increases to 0.986.

8. a. Let A denote the area and w the width, then $A = w(w + 10)$.

b. 20 feet **c.** 50 feet **d.** 120 feet

9. a. She should start descending approximately 34,290 feet,or 6.494 miles, from the airport. She will fly approximately 34,421 feet, or 6.52 miles, before touching down.

b. Approximately 16.7°

10. a. The x-intercepts are 2 and 3. The y-intercept is -6.

b. The equation of its symmetry line is $x = 2.5$.

c. The graph opens down since the coefficient of x^2 is negative.

d. The graphs of $y = -12$ and $y = -x^2 + 5x - 6$ intersect in two points so the equation will have two solutions.

Exercise Set 6, pp. 56–57

1. a.

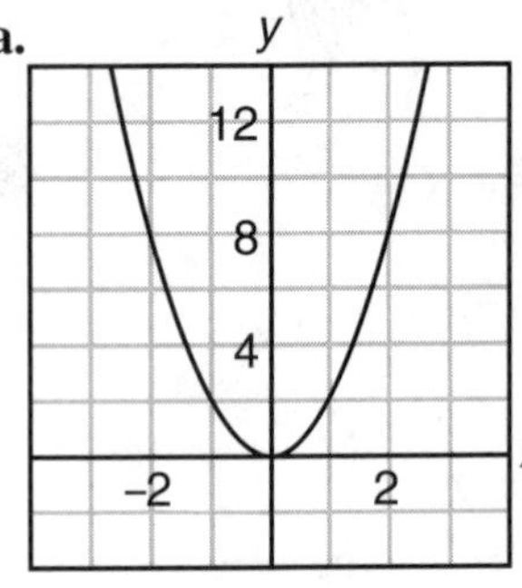

b.

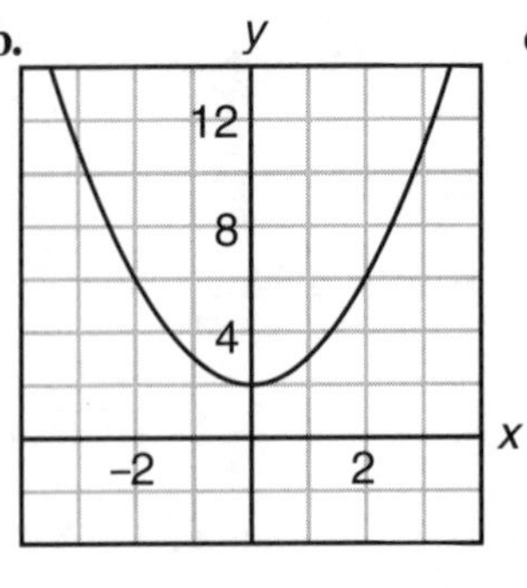

c.

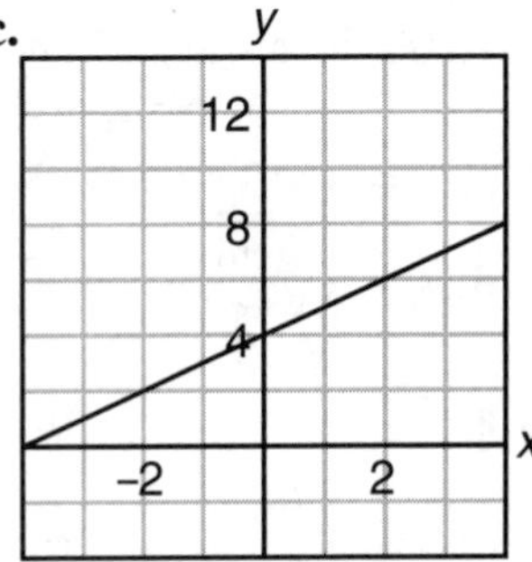

d.

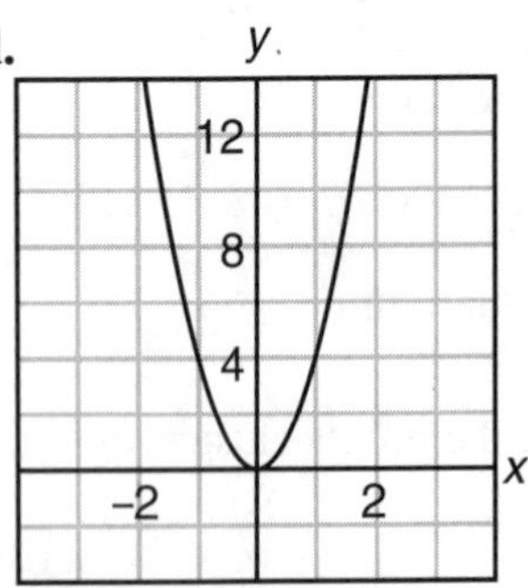

e.

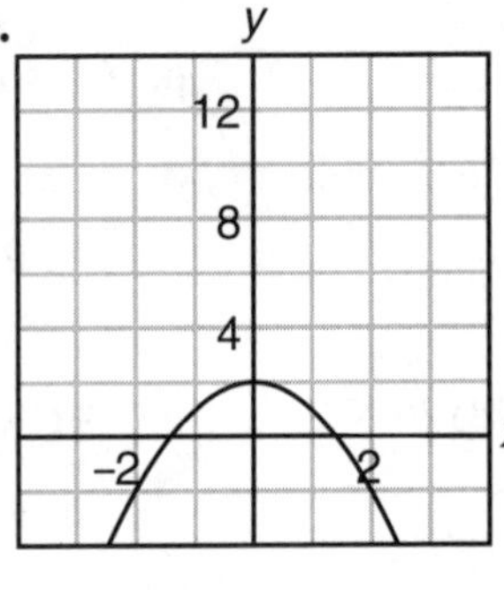

f.

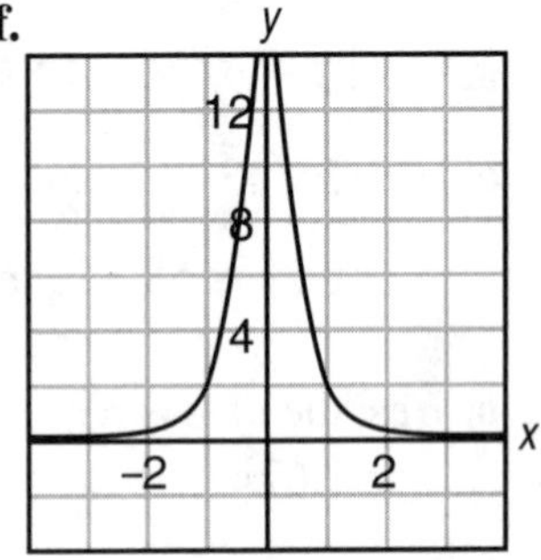

2. $(x, y) = (1.4, 0.2)$

3. a. $x = \pm 2$ **b.** $x = 7$ or $x = -11$ **c.** No solution

d. $x = -2$ **e.** $x = -12$ **f.** $x = \frac{4}{3}$

4. a. $y = 200 + 2.5x$ **b.** $NEXT = NOW + 2.5$ (start at 200)

c. 12 gallons **d.** No more than 20 gallons

e. The company charges for partial gallons on a pro rata basis, rather than rounding up.

5. a. 16 **b.** $8\sqrt{3} \approx 13.86$ **c.** $\frac{6}{7} \approx 0.857$

d. $32x^3$ **e.** $36\sqrt{21} \approx 164.97$ **f.** $72x$

g. $5\sqrt{3} \approx 8.66$ **h.** $2\sqrt[3]{3} \approx 2.88$

6. a. $A'(2, 3)$, $B'(-3, 6)$, $C'(-3, 9)$ **b.** $A'(-2, 3)$, $B'(3, 6)$, $C'(3, 9)$

c. $A'(0, -3)$, $B'(-3, -8)$, $C'(-6, -8)$ **d.** $A'(-12, 8)$, $B'(-24, -12)$, $C'(-36, -12)$

7. a. The x-intercepts are 1 and 10. The y-intercept is 10.

b. The equation of its symmetry line is $x = 5.5$.

c. The graph opens upward since the coefficient of x^2 is positive.

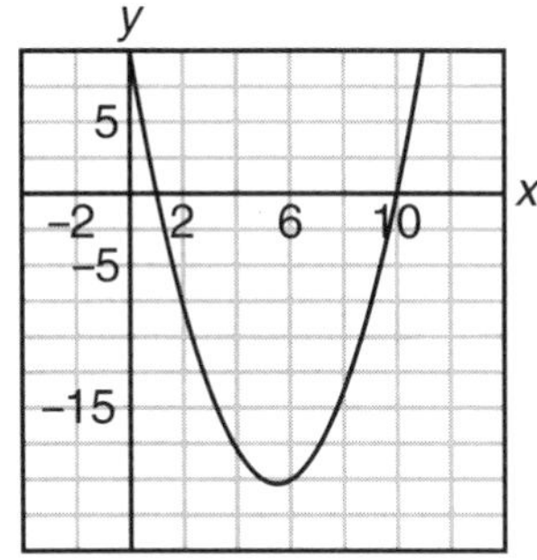

8. a. $c = \sqrt{65}$, $m\angle B \approx 29.74°$, $m\angle A \approx 60.26°$

b. $m\angle P = 55°$, $r \approx 14.34$, $p \approx 20.48$

9. a. $\frac{\pi}{60}$ radians per second

b. $\frac{\pi}{3}$ feet per second

10. 0.55 or 55¢

Exercise Set 7, pp. 58–59

1. a. $y = 10 + 60\left(x - \frac{1}{6}\right) = 60x$

b. $NEXT = NOW + 60$ (Start at 0)

c. 2 hours 5 minutes

d. 70 miles

e. 1 hour 15 minutes

f. 75 miles

2. a. $(x, y) = (3, 5)$

b. $(x, y) = (3, 4)$

c. $(x, y) = (2, 0)$

3. a. $I = \frac{w}{m^2}$, where w is a constant.

b. $w = 20$

c. 3.2 watts per square meter

d. 2.5 meters

4. a. Let p denote the pounds of peanuts and r denote the pounds of raisins. Then $p + r = 3$ pounds.

b. $4.25p + 3.5r = 12$

c. 2 pounds of peanuts and 1 pound of raisins

5. a. $\frac{x^6}{y^4}$

b. x^5y

c. x^6

d. $\frac{162y^2}{x^3}$

e. $\frac{1}{4x^8}$

f. $\frac{18}{x}$

6. a. $C(-8, 3)$ **b.** $AB = \sqrt{10}, BC = \sqrt{10}, AC = 6$

c. $\triangle ABC$ is not a right triangle. If it were a right triangle, it would have to satisfy the Pythagorean Theorem, but it doesn't.

d. $x = -5$

7. a. $\begin{bmatrix} -1 & 5 & 2 \\ 7 & -8 & 2 \end{bmatrix}$ **b.** $\begin{bmatrix} -3 & 1 & 0 \\ 1 & 2 & -2 \end{bmatrix}$ **c.** $\begin{bmatrix} 5 & -1 & 4 \\ -3 & 16 & -1 \\ 0 & -11 & -1 \end{bmatrix}$ **d.** $\begin{bmatrix} 6 & 3 \\ 9 & -6 \\ -9 & 3 \end{bmatrix}$

8. a. \$5.03 **b.** $P = 0.025\pi d^2$ **c.** 12 inch diameter

d. The cost of the 16" pizza is approximately four times the cost of the 8" pizza.

9. Approximately 9.5 meters

10. Approximately 0.331

Exercise Set 8, pp. 60–61

1. a. $y = 25 + 0.40x$ **b.** $NEXT = NOW + 0.40$ (Start at 25) **c.** 186 miles

d. \$43 **e.** 0–175 miles **f.** Up to 250 miles

2. a. $(x, y) = (2, 4)$ or $(x, y) = (-3, 9)$ **b.** $(x, y) = (-1.5, 2.75)$ or $(x, y) = (2, 1)$

c. $(x, y) = (0, 4)$ or $(x, y) = \left(-4\frac{2}{3}, 7\frac{1}{9}\right)$

3. a. $y = -25$ **b.** $x = -8$ and $x = 2$

c.

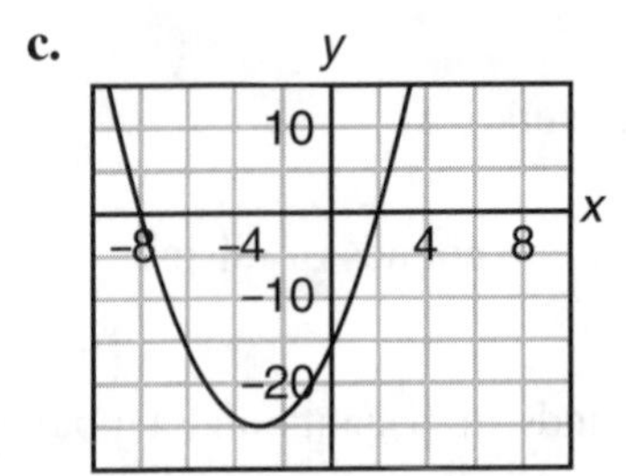

4. a. 91 **b.** $24\sqrt{3}$ **c.** $\frac{2}{3}$ **d.** $18\sqrt{238}$

e. 3 **f.** $2x^6\sqrt{y}$ **g.** $\frac{x}{2y}$ **h.** $x + 7$

5. a. (4, 6)

b. The slopes of $\overline{WX}$ and of $\overline{YZ}$ are -1 and the slopes of $\overline{WZ}$ and of $\overline{XY}$ are 1, so consecutive sides are perpendicular. Thus $WXYZ$ is a rectangle.

c. 16 square units

6. a. 4 centimeters

b. 4.8 centimeters

c.

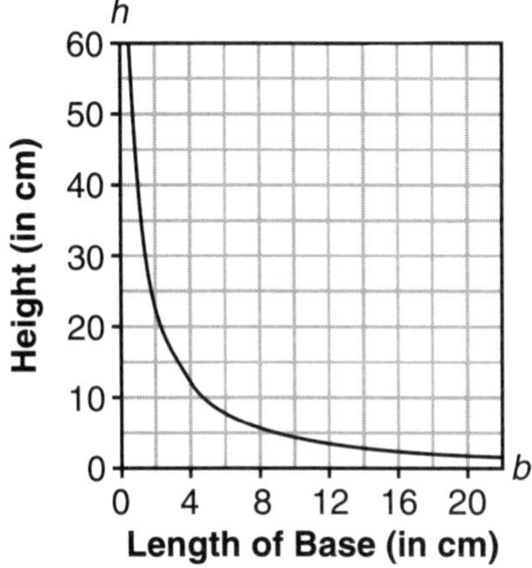

7. a. Approximately 245.04 inches or 20.42 feet

b. $d = 78\pi t$ where d is distance traveled in inches and t is time in seconds.

c. Approximately 14.7 seconds

d. 6π radians per second

8. a. Positive because feet that are wider also tend to be longer.

b. For every inch increase in width, these feet tended to be 2.74 inches longer.

9. a. $\begin{bmatrix} -1 & 0 \\ 0 & 1 \end{bmatrix}$; $(-2, 3)$

b. $\begin{bmatrix} 0.5 & 0 \\ 0 & 0.5 \end{bmatrix}$; $(1, 1.5)$

c. $\begin{bmatrix} 0 & -1 \\ -1 & 0 \end{bmatrix}$; $(-3, -2)$

d. $\begin{bmatrix} 0 & 1 \\ -1 & 0 \end{bmatrix}$; $(3, -2)$

10. a.

$$\begin{array}{c} \\ A \\ B \\ C \\ D \\ E \end{array} \begin{array}{c} \begin{array}{ccccc} A & B & C & D & E \end{array} \\ \begin{bmatrix} 0 & 1 & 1 & 0 & 1 \\ 1 & 0 & 1 & 0 & 1 \\ 1 & 1 & 0 & 1 & 1 \\ 0 & 0 & 1 & 0 & 1 \\ 1 & 1 & 1 & 1 & 0 \end{bmatrix} \end{array}$$

b. 4 paths

Exercise Set 9, pp. 62–63

1. a. $p = 1.2n - 48$

b. 40 notebooks

c. 237 notebooks

2. a. $y = \frac{5}{9}(x - 32)$

b. $y = 50 - x$

c. $y = -\frac{1}{5}(x - 4)$

3. a. $(x, y) = (-1, -2)$ **b.** $(x, y) = (2, -1)$ **c.** No solution

4. a.

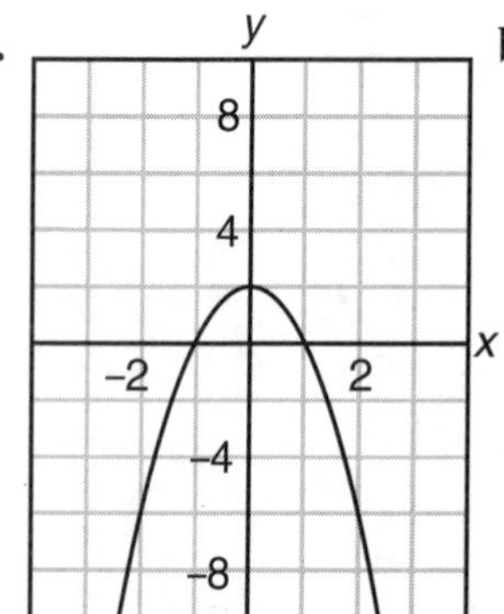

b.

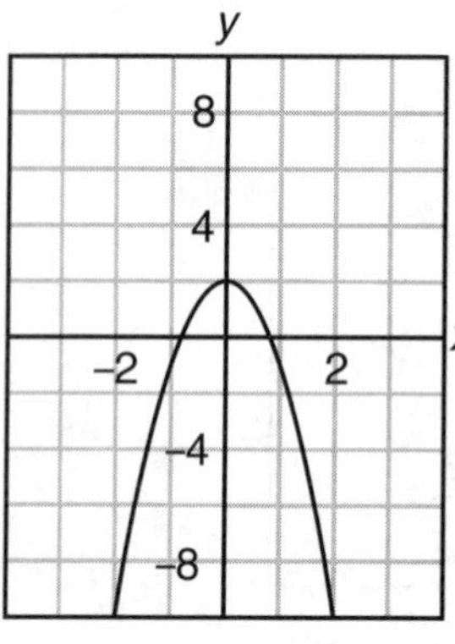

c.

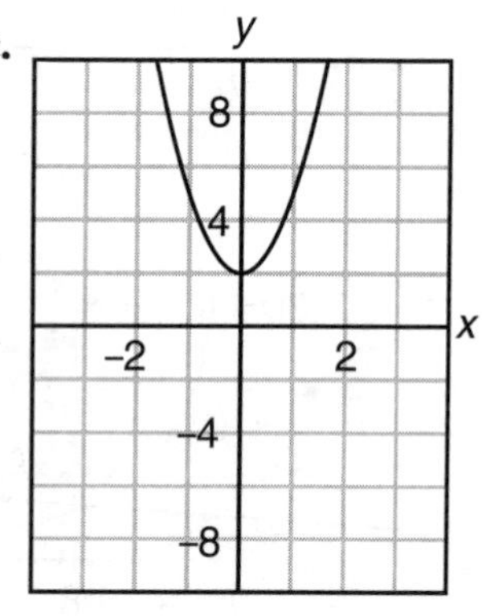

d.

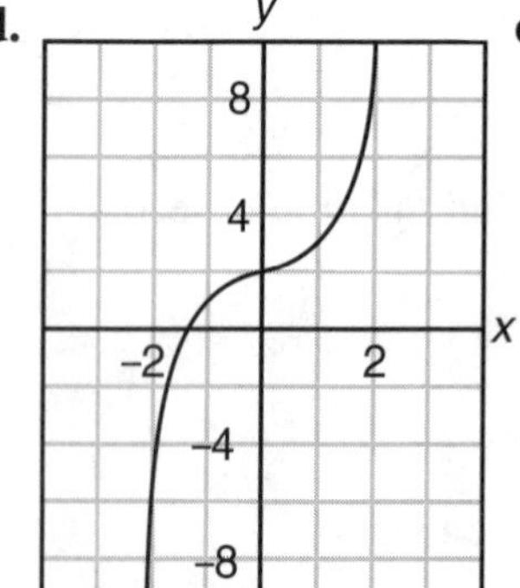

e.

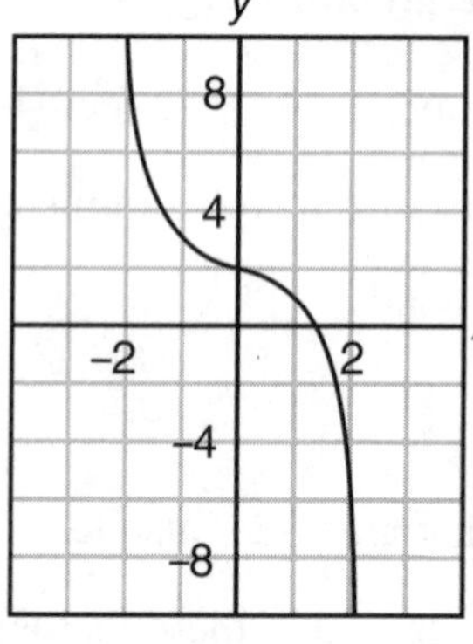

f.

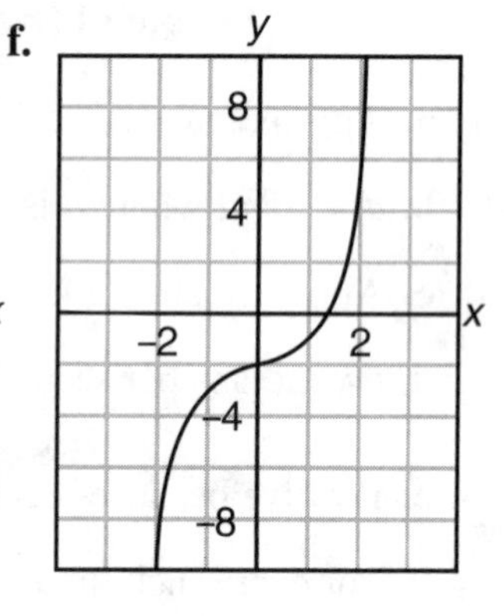

5. a. $x = \pm\sqrt{6}$ **b.** $x = \pm 3$ **c.** $x = \pm 3$

d. $x = \pm\dfrac{\sqrt{30}}{2}$ **e.** $x = \pm\sqrt{\dfrac{37}{3}}$ **f.** $x = \pm 1$

6. a. $x = 7, y = -6, z = 8$ **b.** $x = 2, y = 6$ **c.** $x = 2, y = -4, z = -25$

7. a. The -4.9 represents the downward acceleration due to the force of gravity in meters per second squared; 20 represents the initial upward velocity of the rock in meters per second; and 1.2 indicates that the rock was released 1.2 meters above the ground.

b. Approximately 18.92 meters

c. After approximately 0.50 seconds and 3.58 seconds

d. The rock is at its highest point after approximately 2.04 seconds; its height then is approximately 21.61 meters.

e. Approximately 4.14 seconds

8. a. $A \approx 34.41$ square cm

b. $A \approx 463.64$ square m

c. $A \approx 67.18$ square in.

d. $A \approx 63.59$ square m

9. a. 216

b. 400

c. 13

d. -5

e. 108

f. 2

10. a.

x	1	2	3	4	5	n
$P(x)$	0.4	0.4(0.6)	$0.4(0.6)^2$	$0.4(0.6)^3$	$0.4(0.6)^4$	$0.4(0.6)^{n-1}$

b. $P(4) = 0.4(0.6)^3 = 0.0864$

c. No, since $P(x < 6) = 0.92224$, $P(x \geq 6) = 0.07776$, which is greater than 5%.

Exercise Set 10, pp. 64–65

1. a. $P = 6\sqrt{10} \approx 18.97$ units
$A = 20$ square units

b. $\begin{bmatrix} 0 & 6 & 7 & 1 \\ 0 & 2 & -1 & -3 \end{bmatrix}$;

c. $P = 6\sqrt{10} \approx 18.97$ units
$A = 20$ square units

d. T represents a counterclockwise rotation of 270° or a clockwise rotation of 90° about the origin.

e. $\begin{bmatrix} 1.5 & 0 \\ 0 & 1.5 \end{bmatrix}$

f. $\begin{bmatrix} 0 & -3 & 1.5 & 4.5 \\ 0 & 9 & 10.5 & 1.5 \end{bmatrix}$

g. $P = 9\sqrt{10} \approx 28.46$ units
$A = 45$ square units

2. a. $x = 45$

b. $x = -1$

c. $x = -8$

d. $x = 8$

e. $x = 216$

f. No Solution

3. $(x, y) = (7, 0)$

4. a. y

b. 1

c. $\frac{1}{t}$

d. x^6

e. $\frac{1}{a^6}$

f. $\frac{6}{x}$

g. $\frac{6y^2}{x^2}$

h. $2b^{11}$

5. a. Approximately 4.10 feet

b. Approximately 11.28 feet

c. Approximately 2.74 feet

d. Approximately 7.52 feet

6. a. It is a kite because there are two sets of adjacent sides that are equal.

b. $(-1, -1)$

c. They form a 90° angle since their slopes are 1 and -1.

7. a. Yes, the data is reasonably modeled by a line since the points cluster about a line with negative slope.

b. $y = -1.20x + 8.3$

c. -0.82

d. The residuals are 1.34, 0.24, and 0.96 respectively.

8. a. Let P be the number of pepper plants and s be the space between them. Then $P = \frac{600}{s}$. When $s = 25$ inches, 24 plants fit in a row.

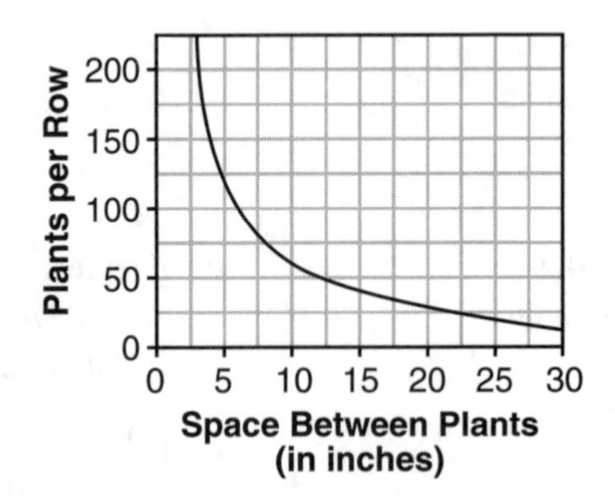

b. Let T be the time to drive between the two cities and s the speed of Nathan's truck, then $T = \frac{165}{s}$. When $s = 60$ mph, it will take 2.75 hours.

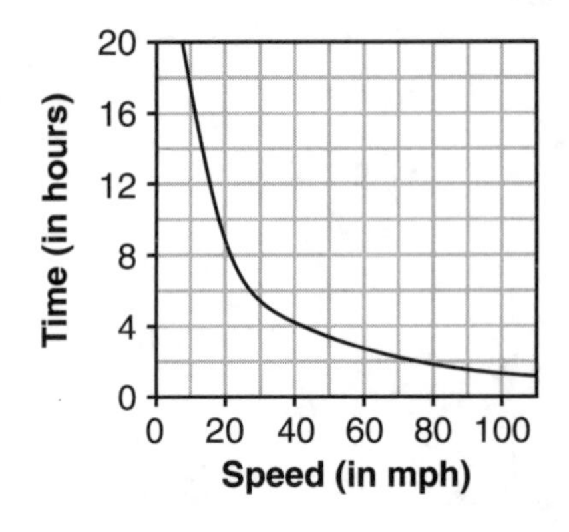

9. a. $\frac{12}{81}$ **b.** $\frac{16}{81}$ **c.** $\frac{9}{81}$ **d.** $\frac{24}{81}$

10. a. $A'(-2, -1), B'(-3, -4), C'(1, -3)$ **b.** $A'(2, -1), B'(3, -4), C'(-1, -3)$

Exercise Set 11, pp. 66–67

1. a. 240 **b.** 45 **c.** 1,000

d. $s = \frac{ru}{kt}$ **e.** $u = \frac{kst}{r}$

2. a. All ordered pairs (x,y) that satisfy the equation $11x - 3y = 100$ **b.** $(x, y) = (37, -61)$ **c.** $(x, y) = (7, 9)$

3. Scatterplots will vary but should have the general distribution shown here.

a.  **b.**  **c.** 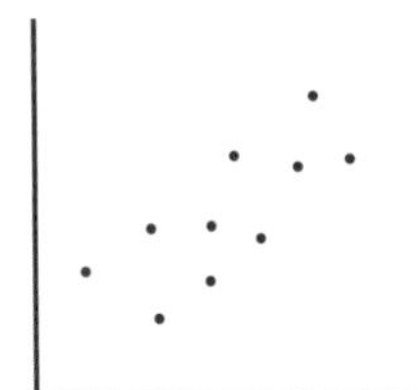

d. 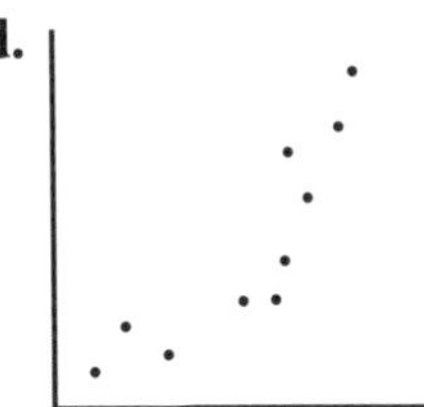 **e.** 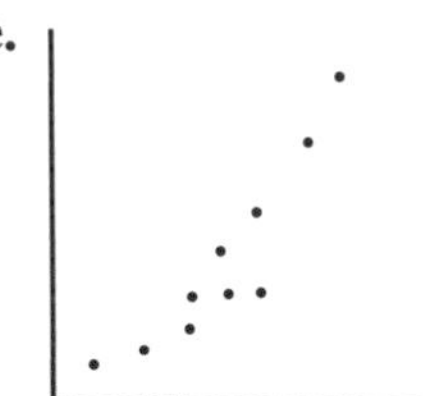

4. a. $x = \pm\sqrt{10}$ **b.** $x = \pm\sqrt{\frac{13}{3}}$ **c.** $x = \pm\frac{\sqrt{39}}{5}$

d. $x = \pm\frac{3}{4}$ **e.** $x = \pm\sqrt{\frac{2}{3}}$ **f.** $x = \pm\sqrt{10}$

5. a. $AB = \begin{bmatrix} -13 & 5 \\ 11 & -5 \end{bmatrix}, BA = \begin{bmatrix} -18 & -1 \\ 10 & 0 \end{bmatrix}; AB \neq BA$

b. $AC = \begin{bmatrix} 1 & -1 & -5 \\ 13 & 47 & 5 \end{bmatrix}$; CA is not defined.

c. $A + B = \begin{bmatrix} 3 & -1 \\ 1 & 4 \end{bmatrix}, B - A = \begin{bmatrix} 7 & -3 \\ -7 & -2 \end{bmatrix}$

d. $\begin{bmatrix} -3 \\ 2 \end{bmatrix}$ **e.** $\begin{bmatrix} -1 & -2 \\ -3 & -5 \end{bmatrix}$

6. a. Domain: All real numbers
Range: $f(x) \geq -9$ **b.** $f(-3) = -5$

c. $x = -8$ or $x = 6$ **d.** $x = -4$ and $x = 2$

7. a. $T = \frac{24}{n}$ where T is the amount of time needed and n is the number of people working.

b.

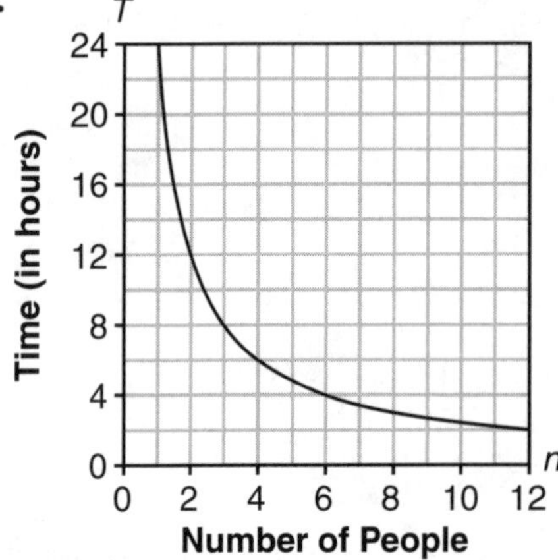

c. 4 hours

d. 16 people

8. a.

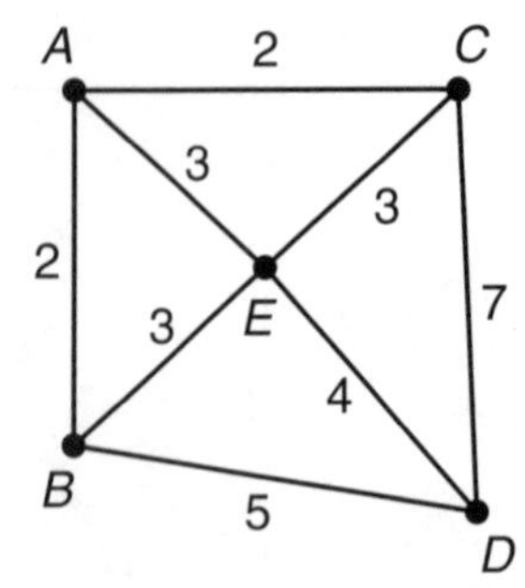

b. One possible minimum spanning tree is shown. The length is 11.

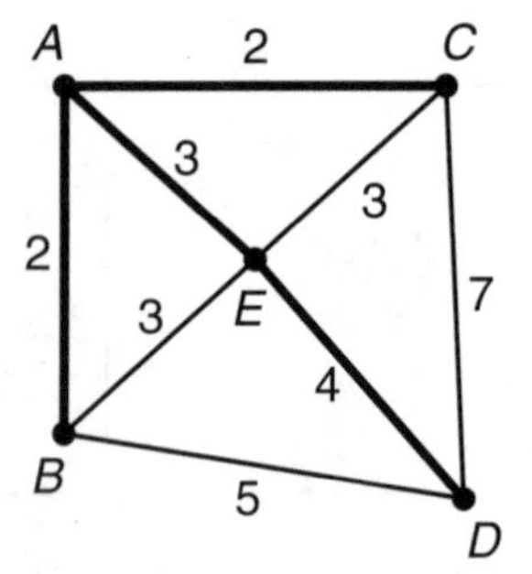

c. The shortest Hamiltonian circuit has a length of 16 and is $A-C-E-D-B-A$.

9. a. $y = 1.28x + 0.5$

b. 0.54

10. a. Approximately 43.56 feet

b. Approximately 33.37 feet

Exercise Set 12, pp. 68–69

1. a. -18

b. 22

c. -928

d. $\dfrac{625}{8}$

e. $-\dfrac{1}{5}$

f. 40

2. a. $h = \dfrac{3V}{\pi r^2}$

b. $r = \sqrt{\dfrac{3V}{\pi h}}$

c. V is tripled.

d. V is multiplied by 16.

3. a.

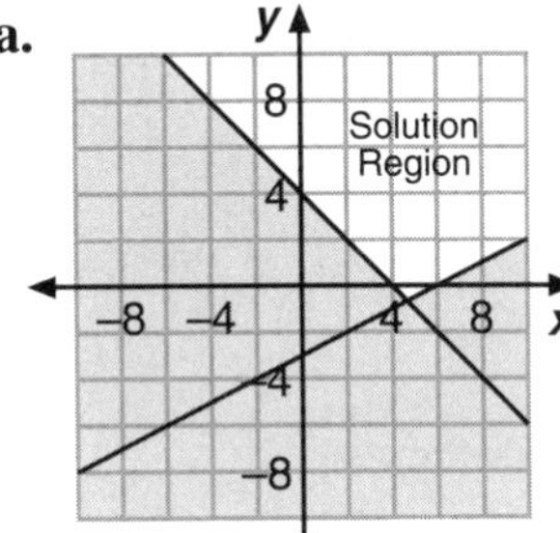

b.

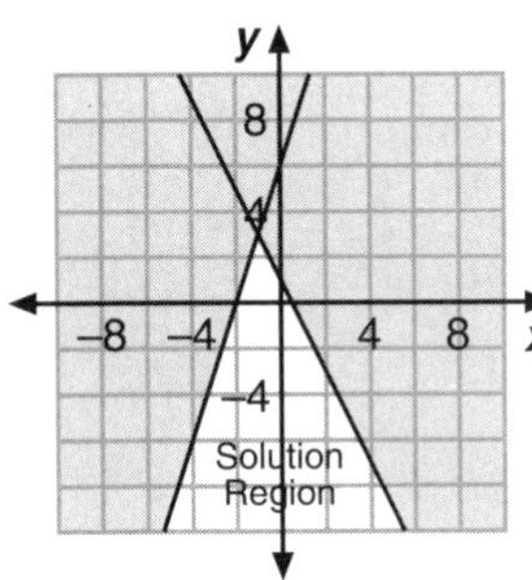

c.

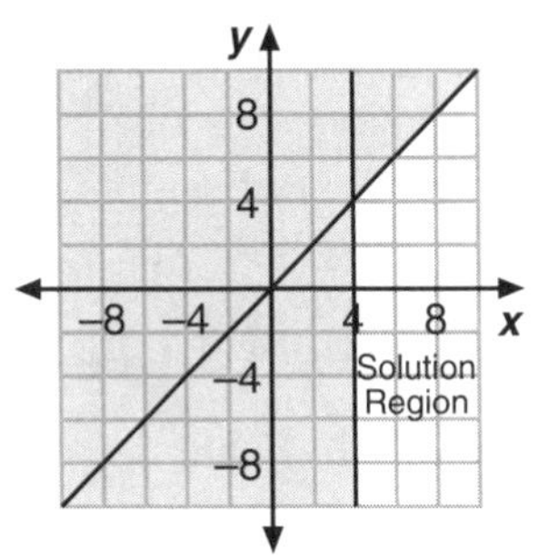

4. a. 18 **b.** 7 **c.** 8 square units

d. Slope of $\overline{CD} = \frac{2}{3}$. Slope of $\overline{BC} = -\frac{3}{2}$.

5. a. $\frac{x}{y}$ **b.** $256x^{12}$ **c.** $-\frac{x^4}{2}$

d. $\frac{4}{x^8}$ **e.** $4x^3y^{10}$ **f.** $\frac{z^9}{y^2}$

6. a. $x = -5$ or $x = 1$ **b.** $x = -1$ or $x = 3$

c. $x = -1$ or $x = 6$ **d.** $x = -5$ or $x = 3$

7. a. Slope of $\overline{BC}$ is -1. **b.** $\triangle ABC$ is isosceles since $AC = AB$.

c. $y = x + 4$ **d.** The point is $\left(-\frac{3}{2}, \frac{5}{2}\right)$. This is the midpoint of $\overline{BC}$.

8. a. $x \approx 5.89$, $y = 8$, $z \approx 79.2°$ **b.** $a \approx 56.25°$, $b \approx 93.82°$, $c \approx 29.93°$

9. a. The x-intercepts are 1 and -5. The y-intercept is -5.

b. The equation of the line of symmetry of the graph is $x = -2$.

c. The graph opens upward since the coefficient of x^2 is positive.

d.

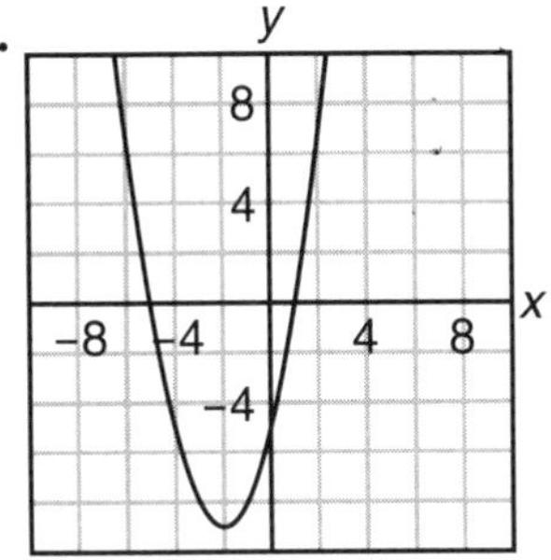

10. No, if you took a random sample of 50 students from this school, getting 26% (13) seniors is a likely sample outcome. The Honor Society doesn't have an unusual number of seniors.

Exercise Set 13, pp. 70–71

1. Graphs a and d can be drawn quickly using x– and y-intercepts

a.

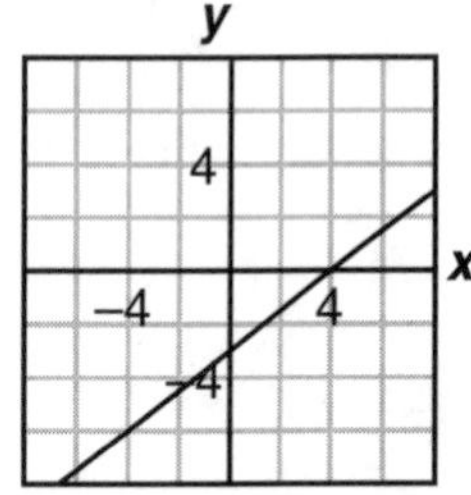

b.

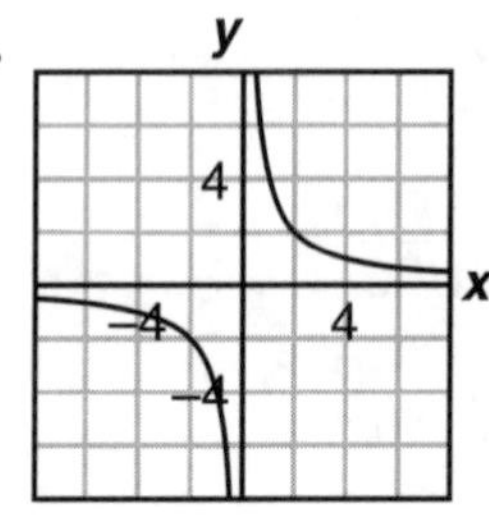

c.

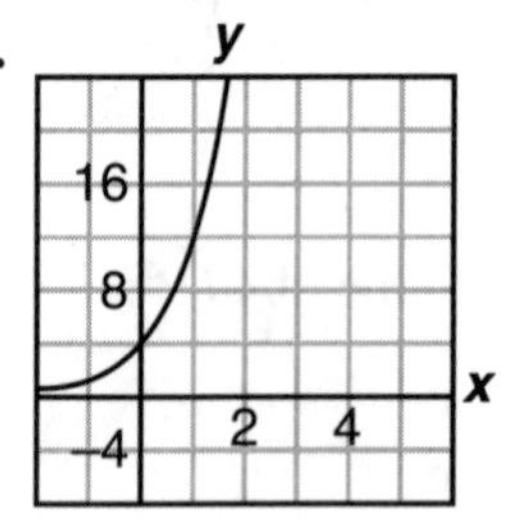

d.

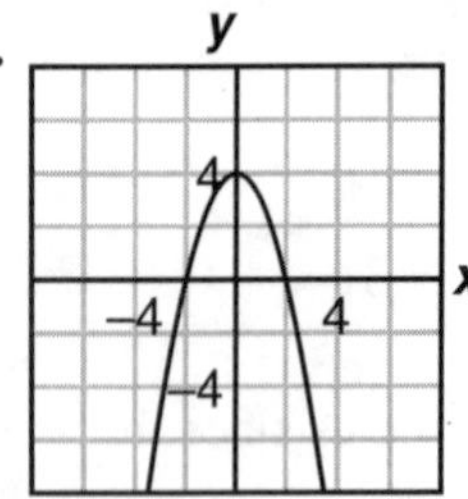

e.

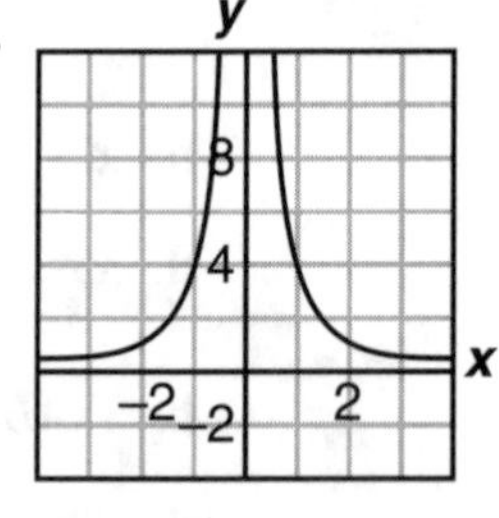

2. a. $x = -12$ **b.** $x = 2$ or $x = -6$ **c.** $x = 2$ or $x = 5$

d. $x = \dfrac{81}{16}$ **e.** $x = -\dfrac{6}{5}$

3. a. $I = \dfrac{120}{R_1} + \dfrac{120}{R_2} + \dfrac{120}{R_3}$ **b.** 10 amps

c. The circuit breaker opens the circuit since the added appliance causes a total current of 16 amps.

4. No. It means that out of every 100 polls taken using the same methods, on average 95 will have a sample proportion that is within 3% of the population percentage.

5. a. $104x^4$ **b.** $24t^2\sqrt{2t}$ **c.** $\dfrac{3x}{4}$ **d.** $2\sqrt[3]{6}$

e. $96\sqrt{43}$ **f.** $3x^4y^2\sqrt[3]{3y^2}$ **g.** $\dfrac{2}{3}$ **h.** $8x^4\sqrt{x}$

6. The midpoint of $\overline{BC}$ is $(-5, -2)$. The equation of the line is $x = -5$. The midpoint of $\overline{AC}$ is $(-3, -2)$. The equation is $y = -\frac{1}{2}x - \frac{7}{2}$. The intersection point is $(-5, -1)$.

7. a. $x = 1$ or $x = 10$ **b.** $x = -14$ or $x = -2$

c. $x = -2$ or $x = 9$ **d.** $x = -2$ or $x = 12$

8. a. Approximately 259.8 feet **b.** Approximately 218 feet

c. Approximately 342.9 feet

9. a. The graph opens downward since the coefficient of x^2 is negative.

b. The equation of the symmetry line of the graph is $x = -1$.

c. $x = 3$ and $x = -5$

d. The maximum value of $f(x)$ is 16.

e.

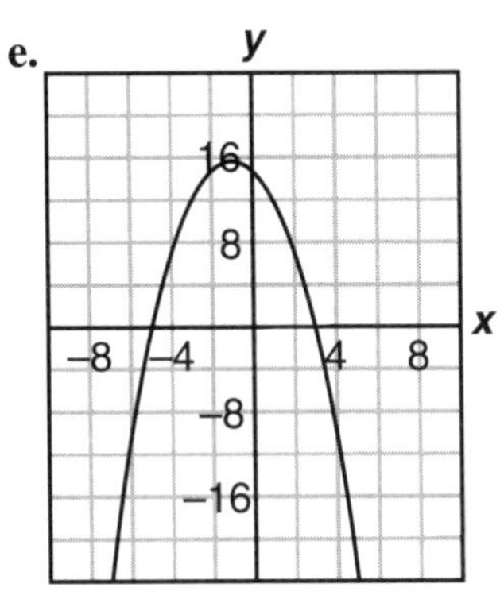

10. a. $\begin{bmatrix} -1 & 0 \\ 0 & 1 \end{bmatrix}$ **b.** $\begin{bmatrix} 5 & 0 \\ 0 & 5 \end{bmatrix}$ **c.** $\begin{bmatrix} 0 & -1 \\ 1 & 0 \end{bmatrix}$ **d.** $\begin{bmatrix} 0 & -1 \\ -1 & 0 \end{bmatrix}$

Exercise Set 14, pp. 72–73

1. a. The coefficient -16 represents the downward acceleration due to the force of gravity in feet per second squared. The 24 represents Carlos' initial upward velocity in feet per second.

b. $f(0.7)$ is Carlos's height after 0.7 seconds. $f(0.7) = 8.96$.

c. Carlos reaches his maximum height after 0.75 seconds. The maximum height is 9 feet.

d. 1.5 sec

2. a.

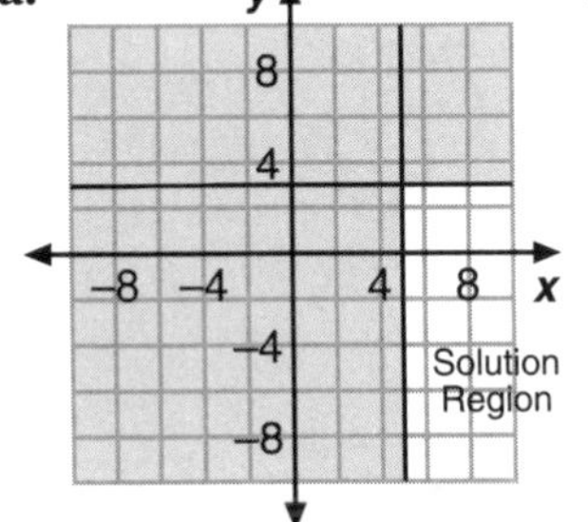

b.

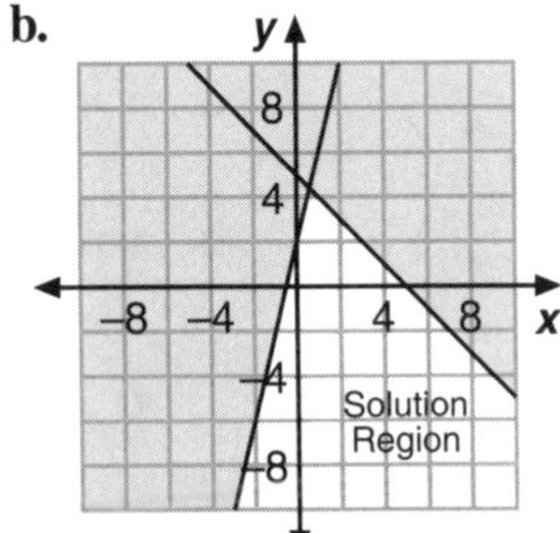

c.

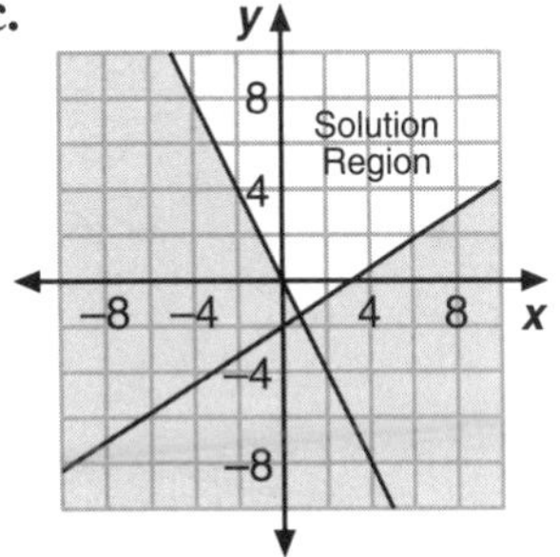

3. a. 44 yards

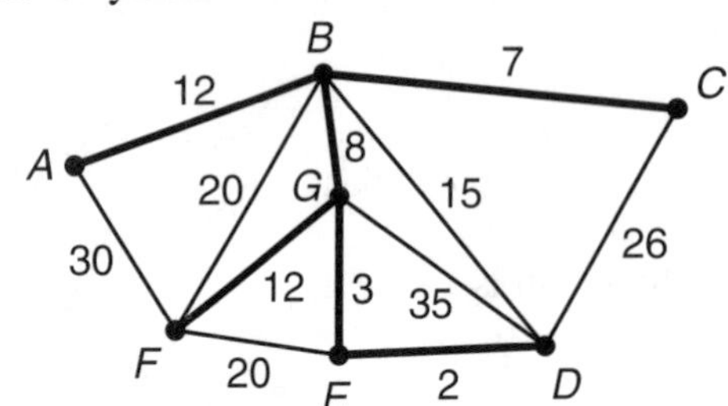

b. 92 yards

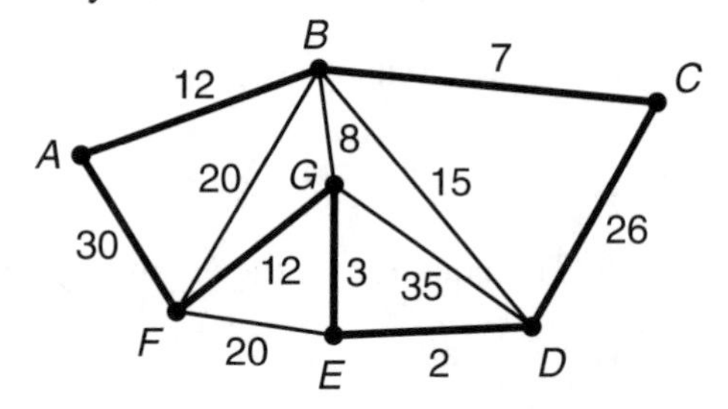

4. $(x, y) \approx (-5.09, 0.36)$

5. a.

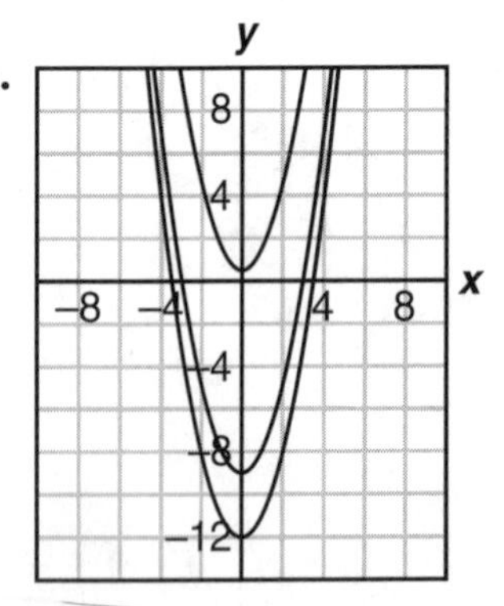

The graphs are congruent but intersect the y-axis in different points.

b.

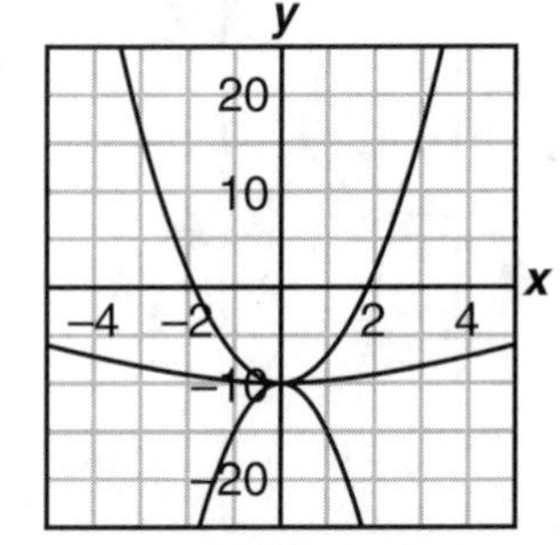

The graphs all have a y-intercept of -10. They differ in width and the direction of opening.

6. a. $5x^2 - 14x$ **b.** x^2 **c.** $x^2 - 4x - 14$

d. $4x^2 + 4x + 1$ **e.** $x^3 + 2x^2 + 2x - 14$ **f.** $-14x - 25$

7. a. It is a parallelogram because opposite sides are parallel.

b. No, the lengths of the diagonals are not equal: $AC = \sqrt{61}$ and $BD = \sqrt{65}$.

c. $\left(-1, -\frac{3}{2}\right)$

8. a. $x = -1 \pm \sqrt{31}$ **b.** $x = 2 \pm \sqrt{2}$

c. $x = -\frac{-5 \pm \sqrt{57}}{-2}$ **d.** $x = 4$ or $x = -1$

9. a. $x \approx 117.3°$, $y \approx 26.4°$ **b.** $x \approx 3.22$, $y \approx 86.5°$

10. a. No. There are two "clumps," each linear on its own, but not linear when combined.

b. Yes. The least squares regression line has the equation $y = 0.765x + 2.64$.

c. The residual for $x = 5.7$ is 0.9995, and the residual for $x = 3.7$ is -1.2705.

Exercise Set 15, pp. 74–75

1. a. $BE \approx 6.84$ cm, $AE \approx 5.32$ cm

b. $DA \approx 7.08$ cm

2. a. $x = \dfrac{b + 7}{a - 3}$

b. $x = \dfrac{c}{2a + 4}$

c. $x = \dfrac{4y + 2k}{y}$

d. $x = \dfrac{ba - 1}{1 - a}$

e. $x = \pm\sqrt{\dfrac{ab}{2}}$

f. $x = k - c^2$

3. The solution set is the unshaded region.

a.

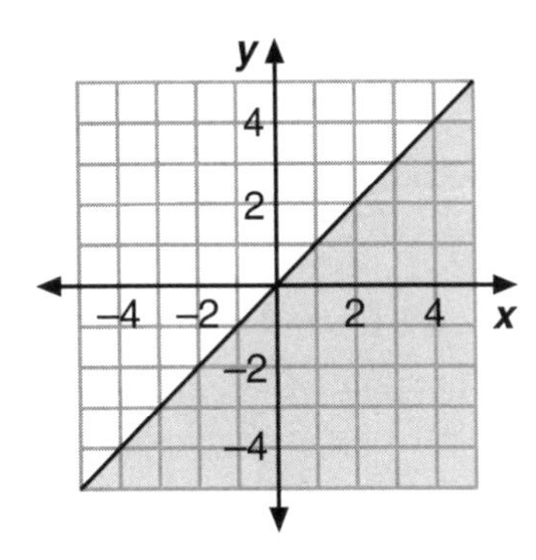

b.

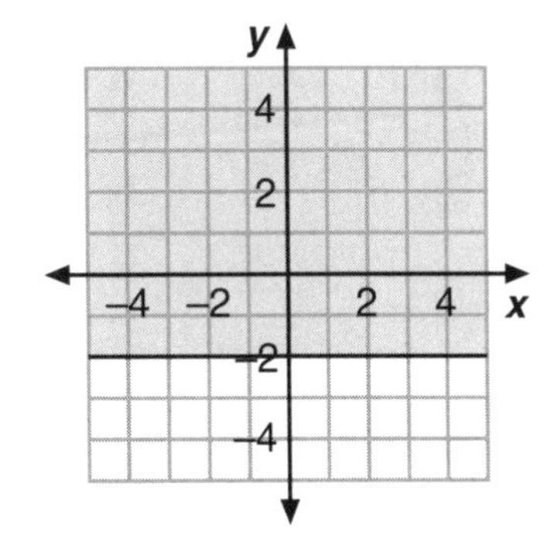

c.

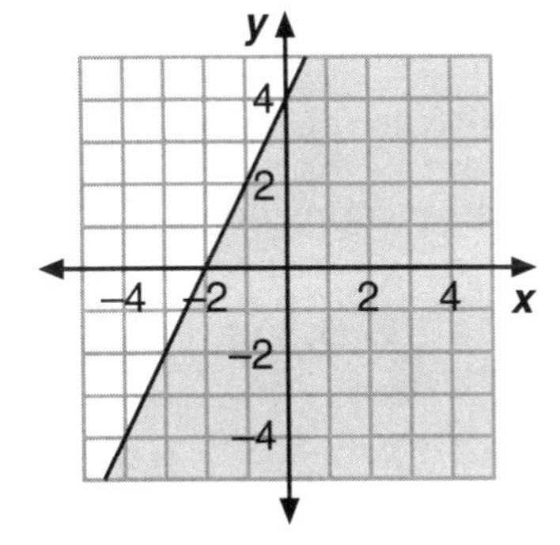

d.

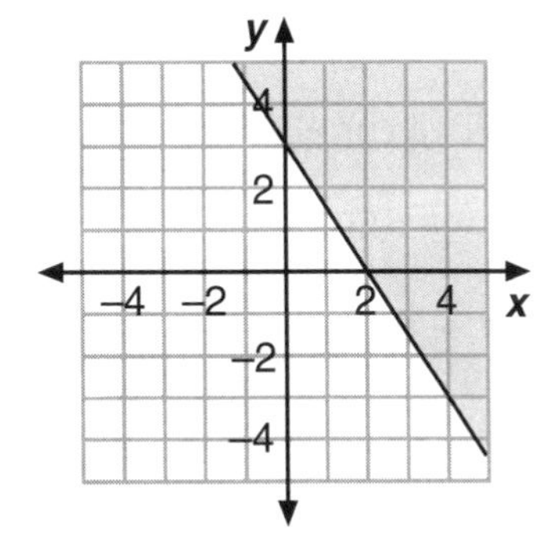

4. 160 adult's and 240 children's tickets were sold.

5. a. $25x^4$

b. $\dfrac{216}{x^3}$

c. $\dfrac{y^6}{x^6}$

d. $\dfrac{3y}{x^2}$

e. $128x^7y$

f. $\dfrac{2}{x^2}$

6. a. The length of $\overline{CA}$ is 5.

b. The length of the segment determined by the midpoints is 2.5.

c. The slopes are equal. The segments are parallel.

7. a. $x = 0$ or $x = 4$

b. $x = 0$ or $x = -3$

c. $x = 0$ or $x = 7$

d. $x = 0$ or $x = 1$

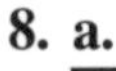

8. a.

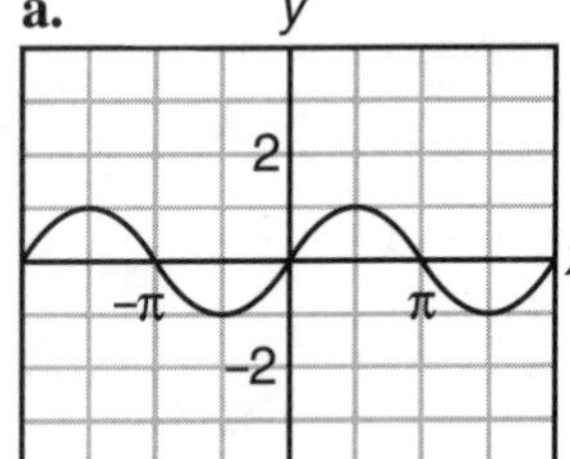

b.

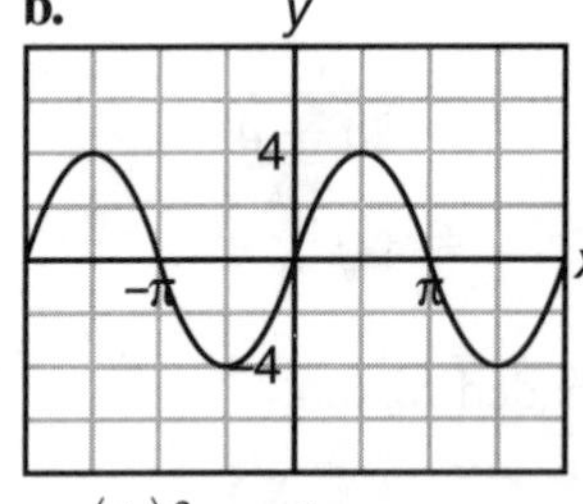

c.

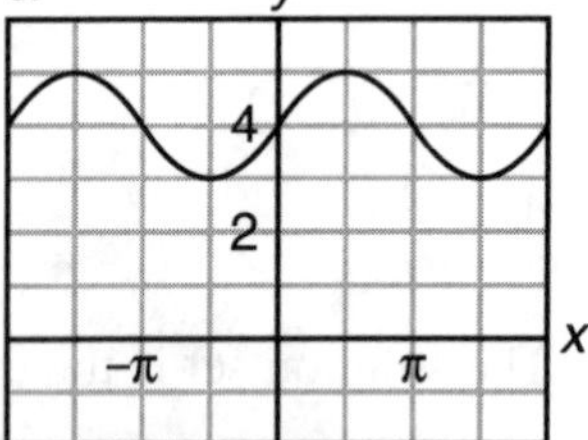

9. a. $\frac{1}{6} \approx 0.167$

b. $\left(\frac{1}{6}\right)^3 = \frac{1}{216} \approx 0.0046$

c. $\left(\frac{5}{6}\right)^5 = \frac{3,125}{7,776} \approx 0.402$

10. a. $x^2 - x - 6$

b. $x^2 - 8x + 16$

c. $x^2 - 25$

d. $3x^2 + 13x - 10$

e. $8x^2 - 14x + 3$

f. $-x^2 + x + 6$

Exercise Set 16, pp. 76–77

1. a. $(x + 6)(x + 2)$

b. $(x + 5)(x + 5)$

c. $(x - 5)(x + 4)$

d. $(x - 6)(x + 6)$

e. $(3x - 1)(x + 4)$

f. $(2x - 1)(3x + 4)$

2. a. Negative correlation

b. Positive correlation

c. Positive correlation

d. Negative correlation

3. a. $(x, y) = (2, 3)$

b. $(x, y) = (2, 7)$

c. $(x, y) = (-1, -1)$

4. a. $x = -\frac{5}{14}$

b. $x = -\frac{17}{4}$

c. $x = 3$

d. $x = \pm 4$

e. $x = -4$ or $x = 3$

f. $x = \frac{34}{3}$

5. 160 crates of apples and 40 crates of pears

6. The statement is true. Since the sum of the angles of a triangle is 180° and the vertex angle has a measure of 60°, the sum of the measures of the two non-vertex angles is 120°. But because these two angles are base angles of an isosceles triangle, they have equal measure. Thus, each has a measure 60°, and all angles are 60°, which implies the triangle is equilateral.

7. a. Domain: All real numbers
Range: $f(x) \geq -49$

b. $f(-3) = -45$
$f(a + 1) = a^2 + 4a - 45$

c. $x = -11$ or $x = 9$

d. $x = -8$ or $x = 6$

e. $f(x)$ has a minimum value of -49.

8. $m\angle 1 = 105°$, $m\angle 2 = 75°$, $m\angle 3 = 105°$, $m\angle 4 = 105°$, $m\angle 5 = 75°$, $m\angle 6 = 75°$, $m\angle 7 = 105°$

9. a. $x = \dfrac{1 \pm \sqrt{13}}{2}$

b. $x = 2 \pm \sqrt{7}$

c. $x = -5$ or $x = 10$

d. $x = -6$ or $x = 7$

10. a. $A'(6, 2)$, $B'(10, -2)$, $C'(-8, -4)$

b. $A'(-1, -3)$, $B'(1, -5)$, $C'(2, 4)$

c. $A'(9, 7)$, $B'(7, 9)$, $C'(6, 0)$

Exercise Set 17, pp. 78–79

1. a. $x > \dfrac{1}{2}$ or $x < -\dfrac{1}{3}$

b. $-\dfrac{1}{3} < x < \dfrac{1}{2}$

c. $x \geq 2$ or $x \leq -\dfrac{5}{3}$

d. $-\dfrac{5}{3} \leq x \leq 2$

e. $x \leq 3$ or $x \geq 4$

f. $-5 \leq x \leq -2$

2. a.

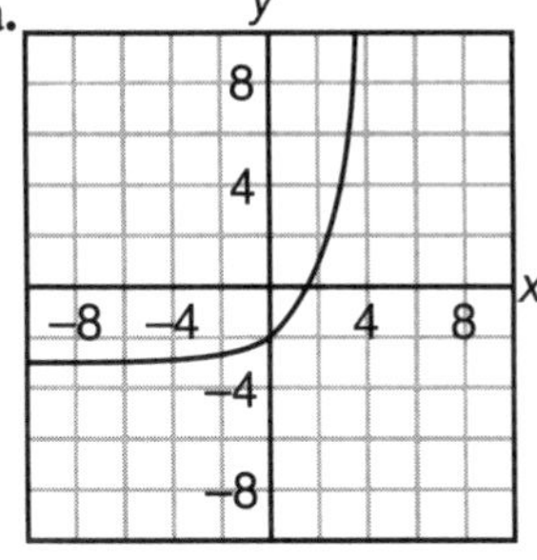

b.

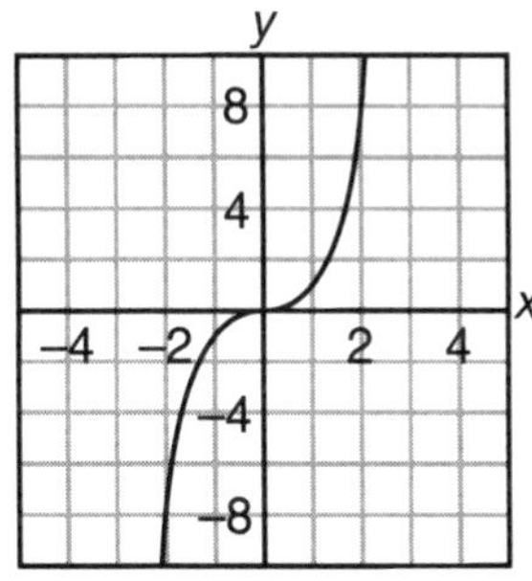

c.

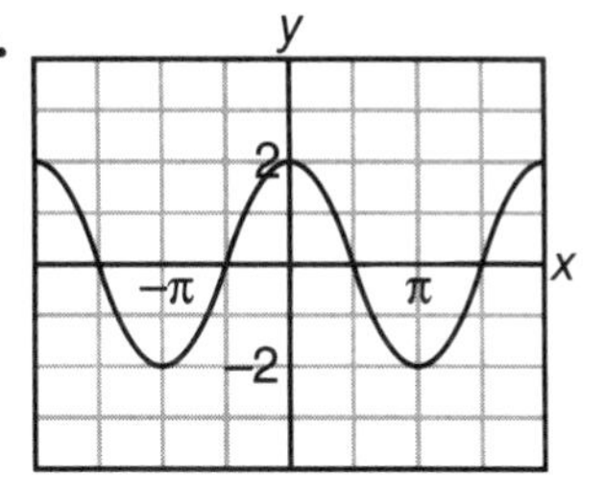

d.

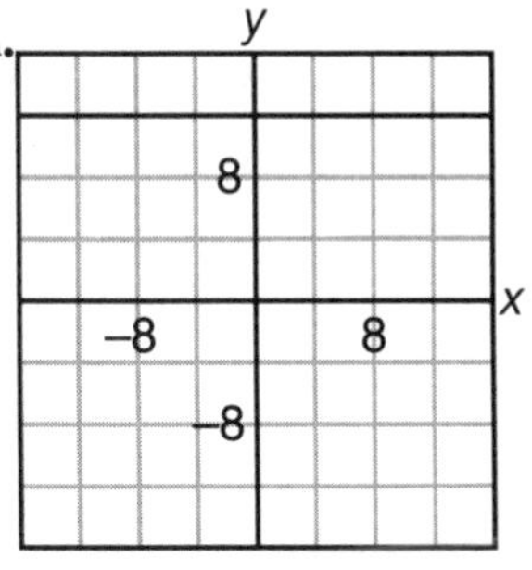

e.

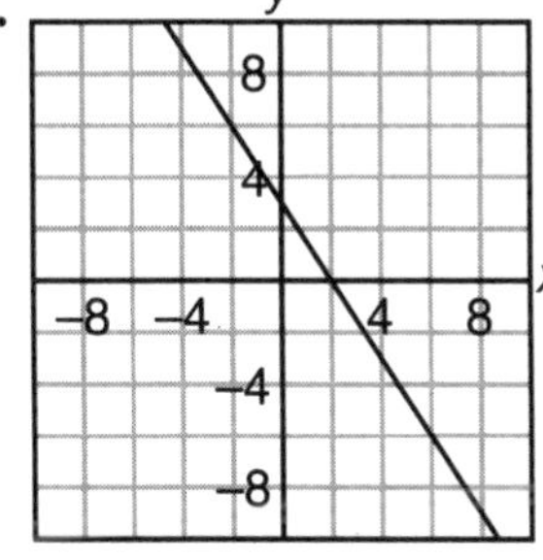

f.

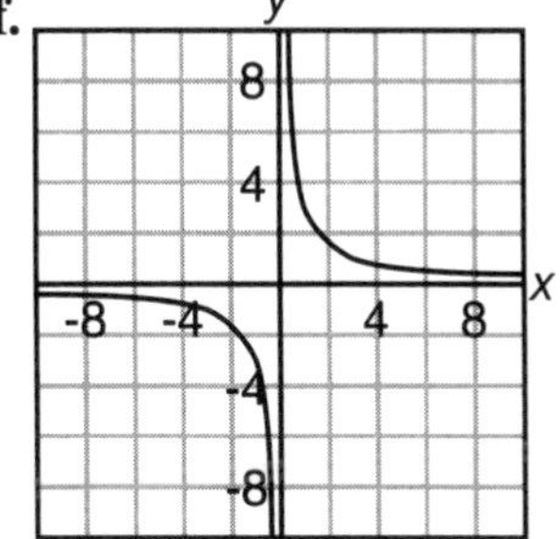

3. a. $3x^2 + 2x + 3$

b. $-3x^2 + 6x - 7$

c. $12x^3 - 14x^2 + 24x - 10$

d. $-9x^2 + 14x - 19$

4. a. x^2

b. y^6

c. r^5

d. $16x^8$

e. $24x^7$

f. $\dfrac{x^8y^2}{4}$

5. a.

$$M = \begin{array}{c} \\ A \\ B \\ C \\ D \end{array} \begin{array}{c} \begin{array}{cccc} A & B & C & D \end{array} \\ \begin{bmatrix} 0 & 1 & 0 & 1 \\ 1 & 0 & 1 & 1 \\ 0 & 1 & 0 & 0 \\ 1 & 1 & 0 & 0 \end{bmatrix} \end{array}$$

An entry of "1" indicates that there is air service between the cities.
An entry of "0" indicates that there is no air service between the cities.

b.

$$M^2 = \begin{bmatrix} 2 & 1 & 1 & 1 \\ 1 & 3 & 0 & 1 \\ 1 & 0 & 1 & 1 \\ 1 & 1 & 1 & 2 \end{bmatrix}$$

Each entry indicates the number of ways to fly between the cities using exactly two flight segments.

c.

$$M^3 = \begin{bmatrix} 2 & 4 & 1 & 3 \\ 4 & 2 & 3 & 4 \\ 1 & 3 & 0 & 1 \\ 3 & 4 & 1 & 2 \end{bmatrix}$$

Each entry indicates the number of ways to fly between the cities using exactly three flight segments.

d.

$$M^0 + M^1 + M^2 = \begin{bmatrix} 3 & 2 & 1 & 2 \\ 2 & 4 & 1 & 2 \\ 1 & 1 & 2 & 1 \\ 2 & 2 & 1 & 3 \end{bmatrix}$$

The absence of zeroes in the sum matrix indicates that you can fly from any of these cities to any other one using no more than two flight segments.

6. a. From the given information, we know that $\frac{AE}{AB} = \frac{2}{3} = \frac{AD}{AC}$. Also $\angle A$ is an angle of $\triangle ABC$ and of $\triangle AED$. Thus, by the SAS similarity theorem, $\triangle ABC \sim \triangle AED$.

b. From Part a, $\triangle ABC \sim \triangle AED$ with a scale factor of $\frac{2}{3}$. Since $\overline{ED}$ and $\overline{BC}$ are corresponding sides of similar triangles, $ED = \frac{2}{3}\,BC$.

c. From Part a, $\triangle ABC \sim \triangle AED$ and, thus, $\angle C \cong \angle EDA$. It follows that $\overleftrightarrow{ED} \parallel \overleftrightarrow{BC}$ since if congruent corresponding angles then the lines are parallel.

7. a. $x = -2 \pm \sqrt{6}$

b. No solution

c. $x = 4$ or $x = -3$

d. $x = 0$ or $x = 8$

8. 0.833 dollars = 83 cents

9. About 1,080 meters

10. a. Yes, the data are reasonably modeled by a line.

b. $y = 1.62x + 7.03$ **c.** $r \approx 0.76$ **d.** -0.046

Exercise Set 18, pp. 80–81

1. a. $(2x - 3)(2x + 3)$ **b.** $(x - 1)(-x + 1)$ **c.** $(2x - 1)(x + 4)$

d. $(2x - 3)(3x - 2)$ **e.** $(2x - 1)(4x + 1)$ **f.** $\left(x + \frac{1}{2}\right)\left(x + \frac{1}{3}\right)$

2. Approximately 2.76 feet from the center of the seesaw

3. a. 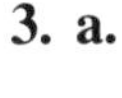

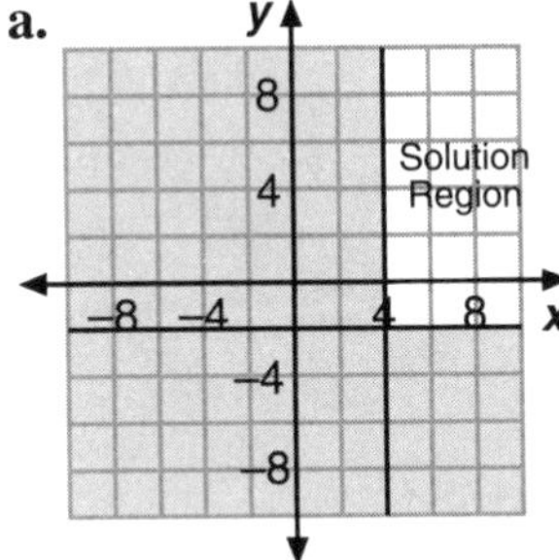

b.

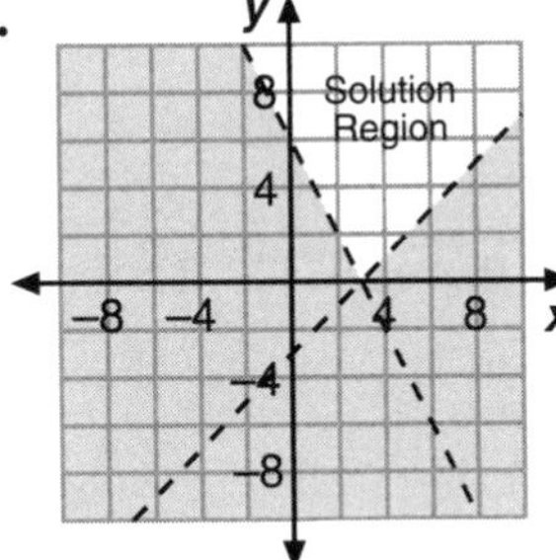

c.

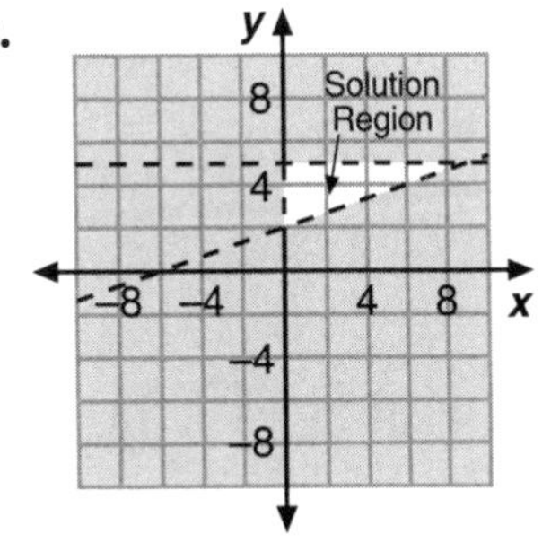

4. a. Yes the data is reasonably modeled by a line.

b. For every additional year of schooling beyond high school, a person's monthly salary in a first job will increase by approximately \$444.

c. Samantha will earn approximately \$888 more than Mia each month.

5. a. $6x^3 - 4x^2 - 2x$ **b.** $16x^2 + 64x + 64$ **c.** $2x^2 + 5x - 12$

d. $12x^2 - 5x - 2$ **e.** $3x^2 + 4x - 4$ **f.** $8x^3 - 4x^2 + 2x - 1$

6. a. $\angle A \cong \angle EBC$ and they are corresponding angles for $\overline{AF}$ and $\overline{BE}$ with transversal $\overleftrightarrow{AB}$. Thus, $\overleftrightarrow{AF} \parallel \overleftrightarrow{BE}$.

b. Since $AB = CD$, $AB + BC = CD + BC$, and thus $AC = BD$. Also, $\angle A \cong \angle EBC$ and $AF = BE$. So, $\triangle AFC \cong \triangle BED$ by SAS.

c. By Part b, $\triangle AFC \cong \triangle BED$. It follows that $\angle F \cong \angle E$ since corresponding angles of congruent triangles are congruent.

7. a. 35 **b.** $32\sqrt{14}$ **c.** $\frac{8}{11}$ **d.** $33\sqrt{93}$

e. $30\sqrt{66}$ **f.** $\frac{2}{3}$ **g.** $\frac{\sqrt{38}}{5}$ **h.** $6\sqrt{3}$

8. a. $(-2\pi, 0)$, $(-\pi, 0)$, $(0, 0)$, $(\pi, 0)$, and $(2\pi, 0)$

b. $(-2\pi, 5)$, $(0, 5)$, and $(2\pi, 5)$

c. $\left(-\frac{3\pi}{2}, 0\right)$, $\left(-\frac{\pi}{2}, 0\right)$, $\left(\frac{\pi}{2}, 0\right)$, and $\left(\frac{3\pi}{2}, 0\right)$

d. $\left(-\frac{\pi}{2}, -3\right)$, and $\left(\frac{3\pi}{2}, -3\right)$

9. a. $x = 4 \pm 4\sqrt{2}$ **b.** $x = \frac{-1 \pm 3\sqrt{5}}{2}$ **c.** $x = \frac{5 \pm \sqrt{65}}{2}$

d. $x = 9$ or $x = -7$ **e.** $x = -4$ or $x = \frac{5}{2}$ **f.** $x = -6$ or $x = -4$

10. a. $\begin{bmatrix} 10 & -1 \\ 2 & 3 \end{bmatrix}$ **b.** $\begin{bmatrix} 0 & 3 \\ 2 & 1 \end{bmatrix}$

c. $\begin{bmatrix} 5 & -2 \\ 0 & -1 \end{bmatrix}$; No, $C \times B \neq B \times C$. **d.** $\begin{bmatrix} 59 \\ 37 \end{bmatrix}$

Exercise Set 19, pp. 82–83

1. a. $p = 0.012w^3$, where p is the power generated in watts and w is the wind speed in mph.

b. The practical domain is probably from 0 to no more than 100 mph. This domain gives a practical range that is greater than 0 up to 12,000 watts.

c. 96 watts **d.** Approximately 25.54 mph

e. The amount of power will be increased by a factor of 8.

2. a. $x = \frac{20}{9}$ **b.** $x = -3$ or $x = -2$ **c.** $x = -\frac{15}{8}$

d. $x = -\frac{5}{3}$ **e.** $x = 2$ **f.** $x = \pm\frac{1}{30}$

3. a. Distributive property

b. Subtraction is inverse of addition.

c. Associative and commutative properties

4. a. $-x^6$ **b.** y^2 **c.** $\frac{x^3y^4}{3}$

d. $-64x^9$ **e.** $12x^{12}$ **f.** $\frac{4x^2}{y^2}$

5. a. $\overline{AB} \cong \overline{CD}$, $\angle BAC \cong \angle DCA$, and $\overline{AC} \cong \overline{CA}$. Thus, $\triangle ABC \cong \triangle CDA$ by the SAS congruence theorem.

b. Since $\triangle ABC \cong \triangle CDA$, $\angle BCA \cong \angle DAC$ since they are corresponding angles of congruent triangles. These angles are alternate interior angles formed by $\overleftrightarrow{BC}$ and $\overleftrightarrow{AD}$ and the transversal $\overleftrightarrow{AC}$ and since they are congruent, $\overleftrightarrow{BC} \parallel \overleftrightarrow{AD}$.

c. Several proofs are possible. One of them is given here. Since alternate interior angles $\angle BAC$ and $\angle DCA$ are congruent, $\overleftrightarrow{AB} \parallel \overleftrightarrow{CD}$. Also $AB = CD$, and so $ABCD$ is a parallelogram because it has a pair of sides that are both parallel and congruent.

6. a. $x = \dfrac{11 \pm \sqrt{41}}{2}$

b. $x = 1 \pm \sqrt{11}$

c. $x = -7$ or $x = 5$

d. $x = -3$ or $x = 7$

7. a. $A \approx 34.47$ square meters

b. $A \approx 46.76$ square centimeters

8. a. 0.30

b. 0.147

c. No. The probability of identifying a student from North Carolina in the first five you stop is 0.83. Thus, the probability of having to stop six or more students is 0.17, which is greater than 0.05.

9. a. Any point R on the y-axis will make $\triangle PQR$ a right triangle. This point will have coordinates $R(0, r)$ for some r.

b. Responses may vary. Any point $R(3, a)$ for some a will work. Additionally, $R(0, 6)$ or $R(0, -6)$, R(6, 6), or R(6, −6) will make $\triangle PQR$ an isosceles triangle.

10. a.

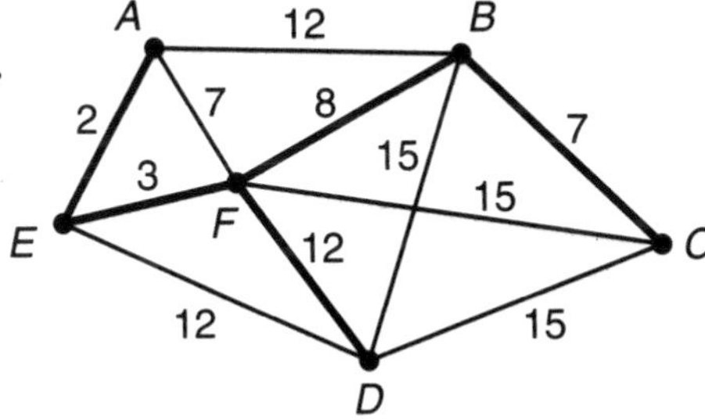

b. 32

c.

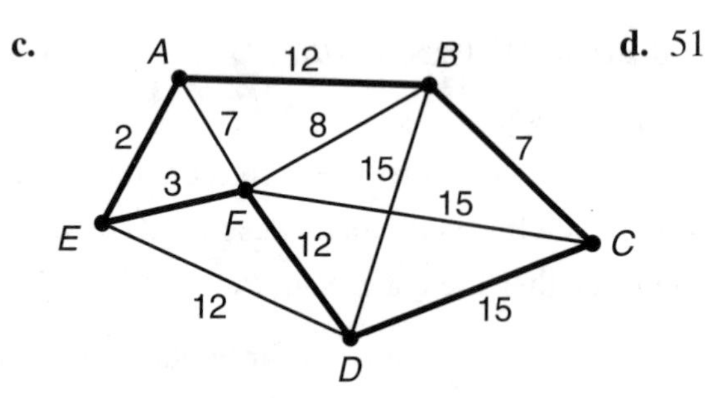

d. 51

Exercise Set 20, pp. 84–85

1. a. $y = 64\left(\frac{1}{2}\right)^x$ **b.** $NEXT = \frac{1}{2}\ NOW$ (Start at 64) **c.** 3

d. 4 inches **e.** 3 or more bounces

2. 35.4°, 85.5°, and 59.1°

3. a. $\frac{3x - 20}{4x}$ **b.** $\frac{x^2 + 6}{3x}$ **c.** $\frac{-x^2 + 9x + 12}{3x}$

d. $\frac{-2x^2 + 3x + 8}{x}$ **e.** $\frac{x + 23}{20}$ **f.** $\frac{5x + 1}{x^2 - 1}$

4. a. $\angle AEB \cong \angle CED$ because they are vertical angles. Since $\angle B \cong \angle D$ it follows that, $\triangle ABE \sim \triangle CDE$ by the AA similarity theorem.

b. Since $m\angle B = m\angle D$ and $\angle B$ and $\angle D$ are alternate interior angles formed by $\overleftrightarrow{AB}$ and $\overleftrightarrow{CD}$ and transversal $\overleftrightarrow{BD}$, $\overleftrightarrow{AB} \parallel \overleftrightarrow{CD}$.

c. Since $\triangle ABE \sim \triangle CDE$, $\frac{AE}{EC}$ is the scale factor, in this case $\frac{6}{3} = 2$. Thus, $\frac{AB}{DC} = 2$ or $AB = 2DC$.

5. a. $x = \pm\sqrt{b - a}$ **b.** $x = \frac{2}{2 - b}$ **c.** $x = \frac{cb - 4a}{a}$

d. $x = (b - 4)^2 - 2a$ **e.** $x = \frac{cb}{ad}$ **f.** $\sqrt[3]{\frac{c - b}{a}}$

6. a. $\frac{1}{12}$ **b.** $\frac{1}{4}$ **c.** $\frac{1}{4}$ **d.** $\frac{1}{2}$

7. a. $r^2 + r - 12$ **b.** $6r^2 + 13r - 5$ **c.** $x^2 - 7x + 12$

d. $-y^2 - 2y + 15$ **e.** $2x^3 - 5x^2 + 5x + 4$ **f.** $9x^2 - 12xt + 4t^2$

8. a. $x < -2$ or $x > 3$ **b.** $-7 \le x \le 4$

c. $x \le -1$ or $x \ge 3$ **d.** $x < -5$ or $x > -3$

9. a. Domain: All real numbers; Range: All real numbers

b. Domain: All real numbers; Range: $f(x) > 0$

c. Domain: $x \geq -2$; Range: $f(x) \geq 0$

d. Domain: All real numbers; Range: All real numbers $g(x) \leq 3$

e. Domain: All real numbers except $x = 1$; Range: All real numbers except $h(x) = 0$

f. Domain: All real numbers; Range: $-5 \leq g(x) \leq 3$

10. a. Approximately 379 feet

b. Approximately 272 feet

Solutions to Practice Sets for Standardized Tests

Practice Set 1, pp. 89–91

1. b
2. e
3. d
4. c
5. c
6. d
7. e
8. b
9. e
10. d

Practice Set 2, pp. 92–95

1. d
2. d
3. a
4. b
5. e
6. c
7. b
8. b
9. c
10. a

Practice Set 3, pp. 96–99

1. c
2. a
3. d
4. e
5. e
6. c
7. d
8. a
9. b
10. a

Practice Set 4, pp. 100–103

1. b
2. a
3. c
4. b
5. b
6. b
7. d
8. c
9. c
10. c

Practice Set 5, pp. 104–107

1. a

2. e

3. d

4. c

5. d

6. e

7. d

8. d

9. e

10. d

Practice Set 6, pp. 108–111

1. b

2. b

3. d

4. e

5. d

6. b

7. b

8. a

9. b

10. e

Practice Set 7, pp. 112–115

1. e

2. b

3. b

4. a

5. b

6. c

7. d

8. e

9. e

10. c

Practice Set 8, pp. 116–119

1. d

2. c

3. b

4. d

5. a

6. c

7. b

8. b

9. e

10. b

Practice Set 9, pp. 120–123

1. **c**
2. **c**
3. **c**
4. **d**
5. **d**
6. **c**
7. **a**
8. **c**
9. **b**
10. **c**

Practice Set 10, pp. 124–128

1. **d**
2. **c**
3. **d**
4. **d**
5. **c**
6. **c**
7. **d**
8. **a**
9. **d**
10. **b**